Informatics and Communication Technologies for Societal Development

Elijah Blessing Rajsingh • Anand Bhojan •
J. Dinesh Peter
Editors

Informatics and Communication Technologies for Societal Development

Proceedings of ICICTS 2014

Springer

Editors
Elijah Blessing Rajsingh
School of Computer Science
and Technology
Karunya University
Coimbatore, Tamil Nadu
India

Anand Bhojan
School of Computing
National University of Singapore
Singapore

J. Dinesh Peter
Department of Information Technology
Karunya University
Coimbatore, Tamil Nadu
India

ISBN 978-81-322-1915-6 ISBN 978-81-322-1916-3 (eBook)
DOI 10.1007/978-81-322-1916-3
Springer New Delhi Heidelberg New York Dordrecht London

Library of Congress Control Number: 2014944381

Printed on acid-free paper

Springer is part of Springer Science+Business Media (www.springer.com)

Preface

This work comprises the proceedings of the International Conference on Informatics and Communication Technologies for Societal Development (ICICTS '14). ICICTS is a joint venture of the professors of the School of Computer Science and Technology of Karunya University and the National University of Singapore. This conference was organized with the primary goal of promoting research ideas that aim to provide solutions to the needs of society through information and communication technologies. ICICTS provided a unique forum for the delegates to exchange ideas, observations, models, results, and experiences with researchers and practitioners who are involved in real-time projects that work towards societal development. ICICTS is supported by a panel of reputed advisory committee members both from India and from abroad. The research track of ICICTS '14 received a total of 101 submissions that were full research papers on diverse fields of information and communication technologies. Each research paper submission was subjected to a rigorous blind review process by three eminent academicians. After the scrutiny, 17 papers were selected for presentation in the conference and are included in this volume.

These proceedings include topics in the fields of image and signal processing, computer vision, communication technologies, soft computing, advanced computing, data mining, and knowledge discovery. The research papers featured in this proceeding provide novel ideas that contribute to the growth of the society through computing technologies. The contents of this proceeding will prove to be an invaluable asset to the researchers in the area of societal development.

The editors appreciate the extensive time and effort put in by all the members of the organizing committee for ensuring a high standard for the papers published in this volume. We would like to express our thanks to the panel of experts who helped us review the papers and assisted us in selecting the candidate for the best paper award. We would like to thank the eminent keynote speakers who have shared their

ideas with the audience and all the researchers and academicians who have contributed their research work, models, and ideas to ICICTS 2014.
Editor

Coimbatore, India	Elijah Blessing Rajsingh
Singapore, Singapore	Anand Bhojan
Coimbatore, India	J. Dinesh Peter

Contents

About the Editors

Elijah Blessing Rajsingh is Professor and Director of the School of Computer Science and Technology at Karunya University, India. He received his M.Tech. and Ph.D. degrees from College of Engineering, Guindy, Anna University, India. His areas of research interest include network security, mobile computing, wireless and ad hoc networks, parallel and distributed computing, grid computing, and pervasive computing. He has several papers to his credit. He is an Associate Editor for *International Journal of Computers and Applications*, Acta Press, Canada, and member of the editorial review board for many peer-reviewed journals. He has been identified as an Expert Member of National Board of Accreditation (NBA), India. His research on health systems is being funded by the Indian Council of Medical Research. He is a member of IEEE, CSI, ISTE and has served as international advisory board member for various international conferences and workshops.

Bhojan Anand is currently working in the School of Computing, National University of Singapore. He received his Ph.D. in Computer Science from the School of Computing, National University of Singapore. His areas of specialization are systems and networking, multimedia systems, and energy management and sustainability. Dr. Anand is very active in the areas of his expertise and has several publications in various reputed international journals. He has authored more than ten books which are widely referred to. He is also an Honorary Secretary of International Researchers Club, Singapore.

J. Dinesh Peter is currently working as Head, Department of Information Technology at Karunya University, Coimbatore. Prior to this he was a full time research scholar at National Institute of Technology, Calicut, India, from where he received his Ph.D. in computer science and engineering. His research focus includes image processing and computer vision. He has published many papers in various international journals and conferences.

Crowd Counting Using Wireless Infrared Distance Sensors for Indoor Environments

Liu Yan, Liu Chao, Peng Ke Wei, and Shao Tao

Abstract Indoor crowd counting has wide applications in various areas, such as security surveillance, estimating customer flows in shopping malls and monitoring occupancy in public transport systems. Many existing counting techniques rely on video cameras or RFID system, which can be too intrusive or too expensive to implement. In this chapter, we introduce a nonintrusive, low-cost, and easy-to-deploy crowd counting method for indoor environments, using off-the-shelf infrared distance sensors and wireless motes. Its nonintrusive nature makes the proposed method more adoptable under circumstances with privacy concerns. The collected signal is processed locally in the wireless motes to reduce the network overhead. The detection result is transmitted back to the base station through the ZigBee wireless radio in the motes. The proposed approach achieved an average detection accuracy of 95 % in the experiments, which shows that it is a highly practical approach for the indoor environments.

Keywords Target detection • Wireless sensor • Data analysis

L. Yan (✉)
Center of Information Technology, Jiujiang University, 551 Qianjin East Rd, Jiujiang, Jiangxi, People's Republic of China
e-mail: liuyan@jju.edu.cn

L. Chao
Institute of Information Science and Technology, Jiujiang University, 551 Qianjin East Rd, Jiujiang, Jiangxi, People's Republic of China
e-mail: liuchao@jju.edu.cn

P. Ke Wei
Jiujiang Vocational University, 91 Jiu Rui Avenue, Jiujiang, Jiangxi, People's Republic of China
e-mail: pkwrick@gmail.com

S. Tao
HP Labs Singapore, 1 Fusionopolis Way #14-07 Connexis, Singapore, Singapore
e-mail: tao.shao@hp.com

E.B. Rajsingh et al. (eds.), *Informatics and Communication Technologies for Societal Development*, DOI 10.1007/978-81-322-1916-3_1, © Springer India 2015

1 Introduction

Indoor crowd counting has exhibited great importance in various areas, such as security surveillance, retail business, and particularly the public transport systems [1, 2]. By monitoring the number of passengers boarding and alighting in real time, the service provider will be able to know the demand for various service routes and the capacity of the vehicles; thus, they plan service schedule accordingly to reduce seat contentions [3]. Many existing counting methods resort to RFID systems for its high accuracy. It is viable in public transport systems where fare cards are widely adopted; however, it can be expensive to implement and also requires cooperation from the monitored objects. For noncooperative scenarios, various image-based counting techniques have been proposed, where cameras are deployed at the entrance/exit areas to keep track of the movements. The targets are detected by analyzing the video frames, imposing a high demand on storage, network bandwidth, and computational power. The performance of image-based methods are constantly affected by the lighting conditions and background patterns, which requires extra tuning when deployed in different environments.

In this chapter, we propose a low-cost indoor crowd counting approach using commodity equipments: infrared distance sensors and wireless motes. The sensors are deployed in pairs to collect distance readings from any objects passing by. The input readings are relayed to tiny wireless motes running the detection algorithm to compute the output event. The final result is transmitted to the base station through the 2.4 GHz ZigBee wireless radio. We conducted actual experiments and collected one week of data to evaluate the performance of the hardware and the detection algorithm. The result shows that the proposed approach can obtain a detection accuracy of 95 % with very low false-positive rate.

The rest of this chapter is organized as follows. The related work is given in Sect. 2. The crowd counting algorithm using the infrared distance sensors is introduced in Sect. 3, and the experiment result is explained in Sect. 4. We conclude in Sect. 5.

2 Related Works

Counting the number of passengers aboard accurately can be a very challenging task in transportation systems when electronic payment systems are not widely adopted. Greneker et al. [4] explored the method of using pressure-sensitive imaging system to detect the number and the moving direction of passengers stepping on the sensor mat based on the pressure pattern. The pressure mat used by this method requires much higher efforts for deployment and maintenance. The problem of counting multiple passengers walking simultaneously remains an unresolved issue. Singapore bus service provider SBS Transit [5] installed infrared sensors on double-deck buses to count the number of available seats in the upper deck. The sensors were mounted on the side of the narrow stairway at the foot level; therefore, the scenario the system faces is simpler compared to monitoring the doorway where a high volume of crowds can move in either direction simultaneously. However, it demonstrates the promise of using infrared sensors for low-cost counting.

Besides sensors, video cameras [6] have also been employed to tackle the people detection problem as it provides more details about the target events, and the spatial correlation [7] among the frames can be exploited for object detection. The Cyclops system [8] used a video image camera to detect passenger flow on trains by analyzing the change in illumination levels caused by passenger movements at the doorway. Although video cameras can be installed easily, video frames must be stored and processed continuously, which can be infeasible when processing capacity, memory space, and network bandwidth are constrained.

3 Proposed Algorithm

In bus systems that provide free ridership to passengers or rely on cash payments, there are no built-in facilities to exploit for tracking of the passengers boarding or alighting buses. The high volume of daily passenger flow also renders any counting method that requires participation of all passengers impractical due to the high level of cooperation required.

As a result, we explored a low-cost, ubiquitous approach that deploys infrared sensors to actively monitor the boarding and alighting events without affecting the regular activity of the passengers. The main components of the counting system are shown in Fig. 1, including a Soekris net5501 embedded computer, Zolertia Z1 motes, and Sharp 2Y0A02 infrared distance sensors.

The Sharp 2Y0A02 infrared distance sensor has an effective range between 20 and 150 cm, which can help to detect the presence of a passenger at the area monitored. We group two infrared sensors as a pair, such that the moving direction of the passenger can be determined based on the fluctuations of the readings from both sensors in the time domain. Each pair of sensors is connected to a Zolertia Z1 wireless mote that processes the sensor readings in real time and reports the entry/exit event detected back to the Z1 mote that acts as a base station. The base station is attached to a Soekris net5501 embedded computer installed on the bus, which will provide the power to the sensor motes through the USB interface, update the onboard passenger count based on the events reported, and send back the occupancy information to the back-end server through the campus wireless network.

The two infrared sensors in a pair are placed 5 cm apart based on empirical results, while the sensing interval is 10 ms, leading to 200 samples per second in total. The handling procedure of the new sensor readings is illustrated in Fig. 2. As the Z1 mote has a limited memory size of 10 KBytes, the buffered readings are stored in an array with $n = 800$ slots or 1.6 KBytes, given that each sample is a 2-byte integer. A threshold value is maintained as the running average of the input

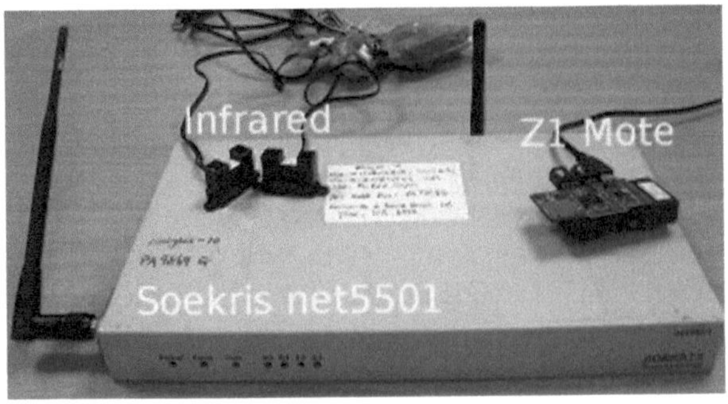

Fig. 1 Devices for infrared-based counting

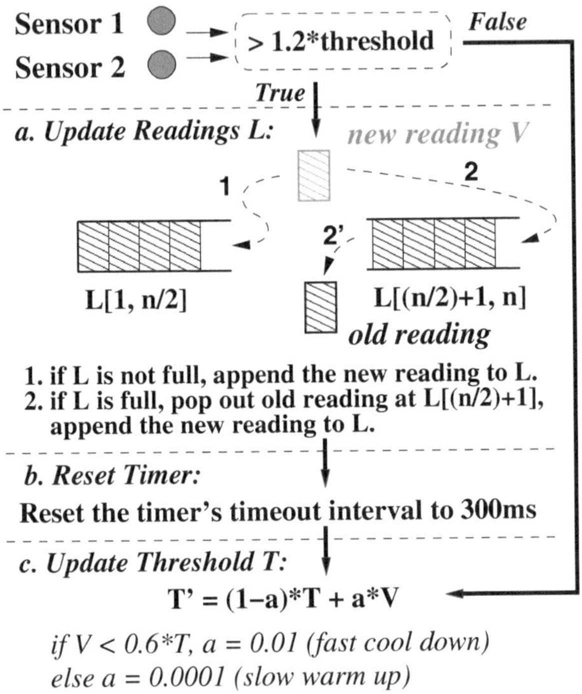

Fig. 2 Handling new sensor readings

readings. Only new readings that are 20 % higher than the threshold are stored to the array, which filters the noise in the data. Given the size of the array, it can capture maximally 4 s of readings in an event. When the sensor is continuously blocked by a lingering passenger, the array may become full. To prevent data overflow, the second half of the array is updated as a FIFO queue, such that the samples in the array cover the first and the last 2 s of the event, which will generally reveal the passenger moving direction.

As the passenger approaches the sensor, the readings will grow above the threshold, and these new readings will be inserted into the array. Each time when

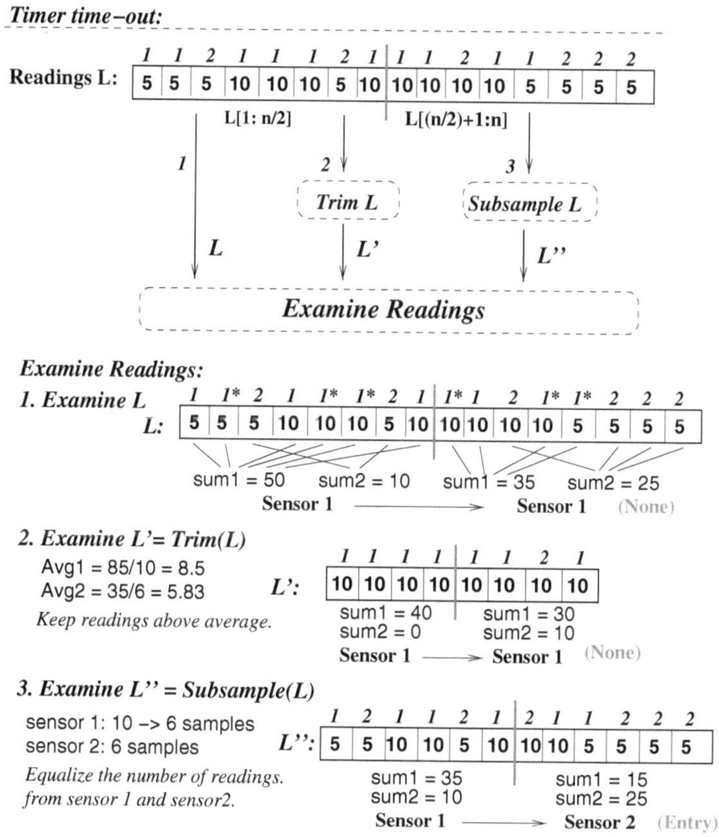

Fig. 3 Examination of readings

a new reading is inserted, an event detection timer is reset to time out after 300 ms. As the passenger moves away from the sensor, the reading values will drop below the threshold level, leading to the event detection timer to expire. Once the timer expires, the event detection method will be called to examine all readings stored in the array and report the type of events detected (*entry* or *exit*). Each sensor maintains its own threshold value, which is updated by each reading as the exponential moving average. For readings less than 0.6 of the threshold, a fast $\alpha = 0.01$ is applied; otherwise, $\alpha = 0.0001$. This introduces a *slow warm-up* and *fast cooldown* effect in the threshold settings. It helps to restore the sensor threshold to nominal level more aggressively after it is blocked by passengers for extended amount of time in crowded scenarios.

The sensor readings stay in the array until the event detection timer times out, after which the data accumulated will be analyzed to generate an event report. The key steps to analyze the sensor readings are depicted in Fig. 3. Let L represent the array and $S1$, $S2$ represent the list of readings in L from sensor 1 and sensor 2. When a passenger boards the bus, sensor 1 will be triggered before sensor 2; therefore, there will be more sensor readings collected from sensor 1 in the initial stage of the activity. As the passenger moves further into the bus, he or she moves away from sensor 1, while sensor 2 will remain active until the passenger leaves the monitored area completely. A similar pattern can be observed when a passenger alights. As these two stages are captured by the first and second halves of L, the type of the event can be determined by calculating the sum of the readings and identify the dominant sensor in $L\left[1:\frac{n}{2}\right]$ and $L\left[\frac{n}{2}+1:n\right]$.

Let $D1$ be the dominant sensor in $L\left[1:\frac{n}{2}\right]$. The sum of all sensor 1 readings in the first half of L is $sum_1 = \sum l_i$, where $l_i \in S1 \cap L\left[1:\frac{n}{2}\right]$. The sum of sensor 2 readings in the first half of L is $sum_2 = \sum l_i$, where $l_i \in S2 \cap L\left[1:\frac{n}{2}\right]$. If $sum1 > sum2$, the dominant sensor $D1$ is *sensor 1*. If $sum1 < sum2$, $D1$ is *sensor 2*. When $sum1 = sum2$, $D1$ is set to null. Likewise, the dominant sensor $D2$ for $L\left[\frac{n}{2}+1:n\right]$ can be obtained. If the transit sequence $D1 \rightarrow D2$ matches *sensor 1* $\rightarrow$ *sensor 2*, the event type is *entry*. The event type is *exit*, if the sequence is *sensor 2* $\rightarrow$ *sensor 1*.

As shown in Fig. 3, examination of the sum values of $S1$ and $S2$ may fail to produce a valid result due to noisy data or overwhelming amount of readings from a single sensor caused by passenger behavior. To counter these issues, we apply two functions *Trim* and *Subsample* to L given sequentially, until a valid result is obtained. The two functions are explained in Algo. 1.

Algorithm 1 *Trim* and *subsample* functions for L

Trim function: Let L' = trim (L).
Let $\bar{l}_1 = \left(\sum_{i=1}^{|S1|} l_i\right)/|S1|$, where $l_i \in S1$.
Let $\bar{l}_2 = \left(\sum_{i=1}^{|S2|} l_i\right)/|S2|$, where $l_i \in S2$.
$\forall l_i \in L$, **if** $(l_i \in S1, l_i > \bar{l}_1)$ or $(l_i \in S2, l_i > \bar{l}_2)$:
    append l_i to L'

Subsample function: Let L'' = subsample (L).
Let $s = min\ (|S1|, |S2|)$, $w1 = s/|S1|$, $w2 = s/|S2|$
Let $sum\_w1 = 0$, $sum\_w2 = 0$
for $l_i \in L$:
    if $l_i \in S1 : sum\_w1 += w1$
        if $sum\_w1 \geq 1$: append l_i to L'', $sum\_w1 -= 1$
    else: $sum\_w2 += w2$
        if $sum\_w2 \geq 1$: append l_i to L'', $sum\_w2 -= 1$

The *Trim* function scans the readings of sensor 1 $S1$ and those of sensor 2 $S2$ in L and computes the average value $\bar{l}_1$ and $\bar{l}_2$ separately. It will remove the readings in L with a value below the average of the corresponding sensors without altering the order of the remaining of the readings values. The result reading array L' will be reexamined for outputs. If L' fails, the *Subsample* function will be applied to L, which will equalize the total number of readings from both sensors by subsampling

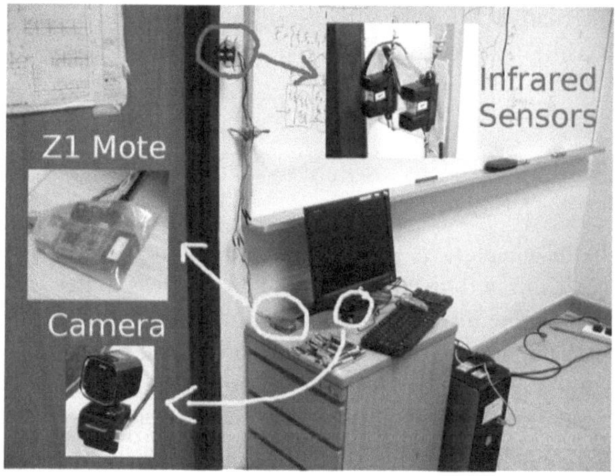

Fig. 4 Sensor deployment for indoor experiments

Table 1 Indoor infrared counting results

Day	1∼2	3∼4	5∼6	7∼8	Sum
Total	448	540	489	415	1892
Correct	422	510	476	389	1797
Wrong	15	17	1	15	48
Miss	11	13	12	11	47
Accuracy	94.20 %	94.44 %	97.34 %	93.73 %	94.98 %

inputs of the sensor generating more readings. The subsampled array L'' will again be examined, and the array L will be cleared after this step.

4 Performance Evaluation

In order to evaluate the infrared counting algorithm, we conducted experiments in a research lab with around 30 students. The infrared sensors were installed close to the entrance horizontally, and a base station mote was attached to a workstation nearby to collect the events reported by the sensors remotely through the wireless radio (Fig. 4). To obtain the ground truth, a web camera was deployed to capture a short video clip of 7 s when the sensors were triggered. The experiment result for an 8-day trial is given in Table 1.

In each day, the number of entry/exit events is around 200∼300 and 95 % of these events can be correctly detected on average. Around 3 % events were classified wrongly, which was mainly due to students standing right in front of the sensors for small talks. Another 2 % of events went undetected, which was caused by multiple students passing through the sensors side by side. The high recall rate of the results shows that the infrared sensor-based method provides a promising alternative for onboard passenger counting on bus systems independent of the existing infrastructure and the service providers.

5 Conclusion

In this chapter, we presented an indoor crowd counting method that utilizes off-the-shelf infrared distance sensors and wireless motes to detect potential target movements. The infrared distance sensors continuously collect the distance readings and pass the input to the attached wireless mote. The wireless mote filters the data input, estimates the movement direction of potential targets, and relays the result to the base station. This scalable design significantly reduces the network bandwidth consumption while allowing multiple locations to be monitored simultaneously for data consolidation. We conducted actual experiments to show that the proposed approach can achieve very high accuracy.

In the future work, we will study alternative deployment orientations of the sensors such that multiple targets passing through the sensors side by side can be identified correctly. We will also explore self-adaptive sensor threshold algorithms such that the false-positive and false-negative rates can be further reduced.

Acknowledgment This work is partially supported by Natural Science Foundation of China under grant No. 61364025, State Key Laboratory of Software Engineering under grant No. SKLSE2012-09-39, and the Science and Technology Foundation of Jiangxi Province, China, under grant GJJ13729 as well.

References

1. SBS Transit. Iris: Intelligent route information system. http://www.sbstransit.com.sg/iris/overview.aspx
2. SMRT Corporation Ltd.: Smrt bus arrival times. http://www.smrt.com.sg/Buses/BusArrivalTimes.aspx
3. Sreedharan, S.: How crowded is the bus today? http://www.todayonline.com/Singapore/EDC120607-0000074/How-crowded-is-the-bus-today
4. Greneker, E., Murphy, K., Johnson, B., Rausch, E.O.: Improved passenger counter and classification system for transit operations. Technical report, Transportation Research Board of The National Academies, Washington D.C. (1996)
5. SBS Transit. SBS transit to introduce Singapore's first wheelchair accessible low-floor superbus. http://www.sbstransit.com.sg/press/2006feb_15-1.aspx (2006)
6. Lienhart, R., Liang, L., Kuranov, A.: A detector tree of boosted classifiers for real-time object detection and tracking. In: Proceedings of ICME, Baltimore, MD (2003)
7. Dollár, P.: Christian Wojek, Bernt Schiele, and Pietro Perona. Pedestrian detection: a benchmark. In: Proceedings of CVPR, Miami, FL (2009)
8. Gene Greneker: Non-contact sensor for passenger counting and classification. Technical report, Transportation Research Board, National Research Council, Washington D.C. (2001)

Target Detection in Infrared Images Using Block-Based Approach

S. Rajkumar and P.V.S.S.R. Chandra Mouli

Abstract A novel and simple method for identification of targets in infrared (IR) images is proposed in this chapter. The input image is divided into blocks of fixed size. The saliency map is extracted block-wise in three different ways. In the first approach, the entropy of each block is compared with the global entropy of the image. In the second approach, the energy of each block is compared with the global energy of the image. In the third approach, combined entropy and energy are used. Experimental results show that the combined approach of entropy and energy detects the targets accurately compared to the other two methods. The subjective results show the efficacy of the proposed approach. The objective evaluation measures show the high detection rate and low false alarm rate.

Keywords Small target detection • Infrared imagery • Entropy-based approach • Energy-based approach

1 Introduction

The advent of IR imaging technology finds mammoth applications in defense and military problems. Target detection is the first and the foremost application of defense over the IR imagery. Even though the target detection in general is a classical and old problem, in the IR domain, it is still challenging. Human vision is limited to a small portion in the electromagnetic spectrum in the form of visible light. The infrared radiation is invisible to the human eye but can sense the heat-emitting objects at any time. Infrared imaging is mainly useful for military- and civilian-related application purposes. Military application includes surveillance, homing and tracking, target acquisition, and night vision, while nonmilitary

S. Rajkumar (✉) • P.V.S.S.R. Chandra Mouli
School of Computing Science and Engineering, VIT University, Vellore, TN, India
e-mail: rajkumars@vit.ac.in; chandramouli@vit.ac.in

E.B. Rajsingh et al. (eds.), *Informatics and Communication Technologies for Societal Development*, DOI 10.1007/978-81-322-1916-3_2, © Springer India 2015

application includes remote temperature sensing, thermal efficiency analysis, weather forecasting, spectroscopy, and environmental monitoring [1].

Infrared images play an important role in military and civil fields for small target detection since 1910 because targets have a low signal-to-noise ratio (SNR) [2–4]. Unlike visible cameras, IR images capture only the heat emitted from objects. The quality of these objects degrades due to sensor noise and other clutter noises. In view of these reasons, the target recognition in IR images is very difficult. Since the target recognition in IR images is difficult, it is required to detect the targets first for target localization.

Target detection in IR images is challenging and has profound applications in defense-related areas. The surveillance aircrafts equipped with IR sensors capture the scenes in field from faraway distances, since the targets in IR images are far away from imaging equipment and hence usually very dim and embedded in clutter background [5, 6]. Shaik and Iftekharuddin [7] have contributed much in the IR imaging domain. They explored the frequency domain correlation and Bayesian probabilistic techniques for the automatic target detection.

Small target detection using two-dimensional least mean square (TDLMS) [8] is a general adaptive filter algorithm applied to infrared small target detection. In this chapter, a new TDLMS filter structure is defined based on neighborhood analysis. Sungho Kim et al. [9] proposed a novel human visual system (HVS) contrast mechanism-based detecting algorithm, which is capable of increasing target intensity as well as suppressing background clutter and noise. In [10], an improved algorithm based on HVS for infrared small target detection in an image with complicated background is discussed. Targets under sea-sky background are detected using entropy and empirical mode decomposition in [11]. Fractal-based methods are used in [12] for target detection. A new concept called self-information map is introduced by Deng et al. in [13] which is integrated with adaptive thresholding for small target detection. Recently, the saliency-based work is progressing for target detection in IR images. Saliency extraction became an important step in the segmentation of images and in object recognition. Visual attention analysis and saliency extraction are applied in [14, 15]. Zhao et al. [16] proposed a real-time automatic target detection using saliency extraction and morphological theory. Local frequency-tuned-based saliency extraction method is proposed in this chapter. The potential targets are obtained by comparing the saliency extracted with adaptive thresholding map. Noise is suppressed as the post-processing step to remove any impulse present. In this chapter, an adaptive target detection method is proposed using block processing of IR image. The proposed method is a modified version of [16]. The subjective and objective evaluation shows that the proposed method detects the objects better than the existing methods.

The rest of this chapter is organized as follows. The proposed methodology is introduced in Sect. 2. Experimental results are presented in Sect. 3. In Sect. 4, performance analysis of the results is done by comparing with existing methods. Finally, Sect. 5 concludes the work.

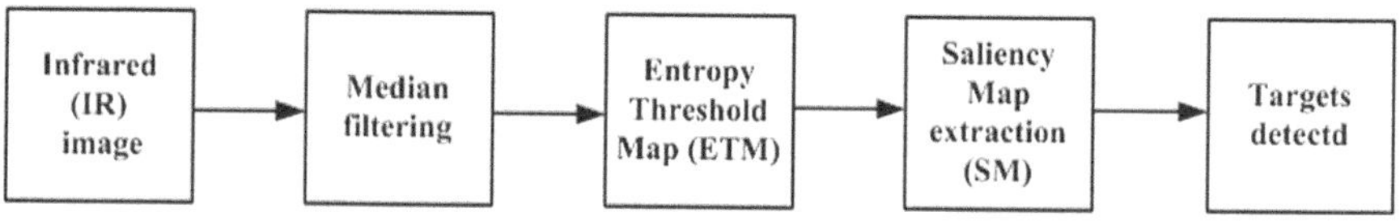

Fig. 1 Block diagram of the proposed method

2 Proposed Methodology

In this section, the proposed methodology is discussed in detail. The block diagram of the proposed method is given in Fig. 1. The IR image is captured from IR sensors from far-off distance and noise is inevitable. To remove the noise, median filtering is applied so that the image is blurred.

2.1 *Saliency Map Extraction*

The median-filtered image is used to extract the saliency map. To do this, block processing of the image is employed. The image is divided into blocks with necessary zero padding. After a series of experiments, the size of each block is fixed as 15×15 for which the detection rate is good. The lesser the block size, the smaller the probability of detection. Similarly, the higher the block size, the higher the probability of false alarm rate.

2.1.1 Entropy-Based Saliency Extraction

The entropy of the median-filtered image is calculated and compared with the local entropy of each block. The salience map is defined as

if (lent > gent) image_block = 1; else image_block = 0; end

where *lent* is the local entropy of the image _ block of size 15×15 and *gent* is the global entropy of the image. The analogy is that if the local entropy exceeds the global entropy, the probability that the block contains the heat-emitting object, i.e., target, is high. The saliency map thus obtained is a binary image where each white block represents the target and the rest is merged with the background.

2.1.2 Energy-Based Saliency Extraction

Instead of entropy, the energy of the median-filtered image is used to extract the saliency map. The energy-based salience map is defined as

if (leng > geng) image_block = 1; else image_block = 0; end

Table 1 OSU Thermal Pedestrian Database details

Collection #	No. of images
1	31
2	28
3	23
4	18
5	23
6	18
7	22
8	24
9	73
10	24

where *leng* is the local energy of the image _ block of size 15 × 15 and *geng* is the global energy of the image.

2.1.3 Entropy- and Energy-Based Saliency Extraction

In this approach, both entropy and energy together are used to extract the saliency map. The saliency map thus obtained is defined as

if ((lent > gent) & (leng > geng)) image_block = 1; else image_block = 0; end

3 Experimental Results

The proposed method is implemented over the OTCBVS infrared image database [17]. The results presented in this chapter are from the OSU Thermal Pedestrian Database. It contains 10 different collections of images. The total number of images in this collection is 284 in which 155 pedestrians (targets) are available. The details of each collection are given in Table 1.

The original image and the corresponding saliency map obtained are shown in Fig. 2. In Fig. 2, the first column represents the original image, the second column represents the saliency map obtained, and the third column represents the labelled map. The labelled map is generated for objective evaluation. The number of connected components is determined from the saliency map. Using the number of connected components, each component is labelled and assigned a unique color for discrimination. From the saliency map, the location and position of targets are determined and superimposed on the border of the target detected over the input image. The results are shown in Fig. 3.

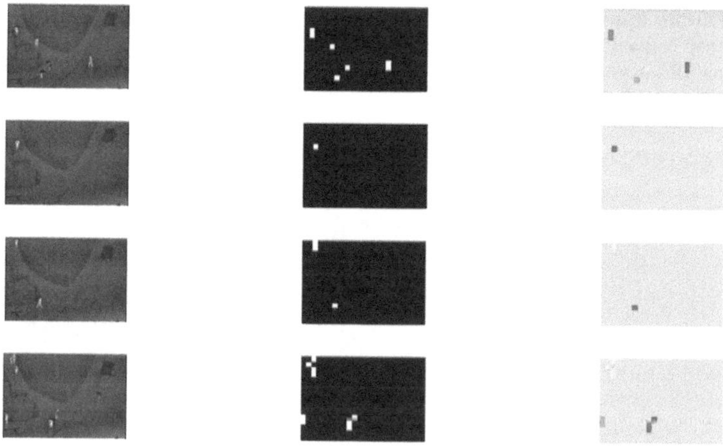

Fig. 2 Saliency map and labelled map obtained for the original IR image

4 Performance Analysis

The results obtained are statistically evaluated using the measures *PD* (detection rate) and *PFA* (false alarm rate). These are defined in Eqs. 1 and 2, respectively:

$$PD = \frac{\text{No.of true targets detected}}{\text{Total no.of targets}} \times 100\% \tag{1}$$

$$PFA = \frac{\text{No.of false targets detected}}{\text{Total no.of targets}} \times 100\% \tag{2}$$

The plots in Fig. 4 show the plot of mean detection rate and mean false alarm rate. From Fig. 4, the combination of entropy- and energy-based saliency map has higher detection rate and lower false alarm rate.

5 Conclusion

A set of three approaches are proposed for target detection in IR images. The proposed approaches are trivial but due to block processing of images help in higher detection rate and lower alarm rate. The combination of entropy and energy parameters led to an aggregate of around 96 %. It is planned to improve the work towards 100 % detection rate. To do that, a robust saliency map is planned by extracting robust features other than entropy and energy as future work.

Fig. 3 Saliency map and labelled map obtained for the original IR image

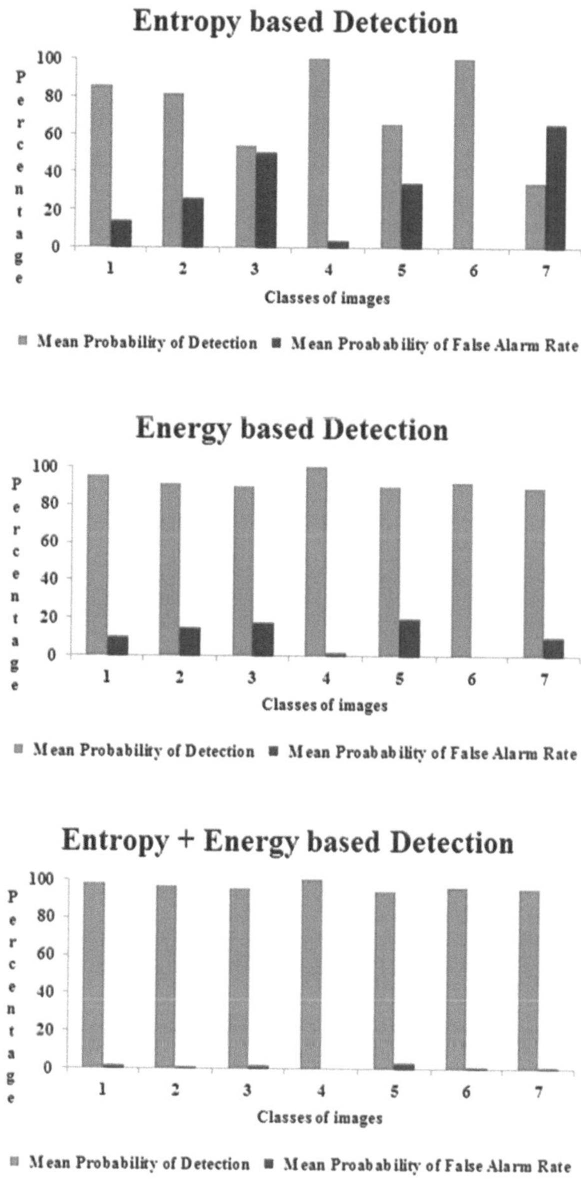

Fig. 4 Detection rate and false alarm rate over the three proposed approaches

Acknowledgments This work is supported by the Defence Research and Development Organization (DRDO), New Delhi India, funding the project under the Directorate of Extramural Research & Intellectual Property Rights (ER & IPR) No. ERIP/ER/1103978/M/01/1347 dated July 28, 2011.

References

1. Russ, J.C.: The Image Processing Handbook. CRC Press, Boca Raton (2006)
2. Bai, X., Zhou, F.: Analysis of new top-hat transformation and the application for infrared dim small target detection. Pattern Recognit. **43**(6), 2145–2156 (2010)
3. Bai, X., Zhou, F.: Hit-or-miss transform based infrared dim small target enhancement. Opt. Laser Technol. **43**(7), 1084–1090 (2011)
4. Peregrina-Barreto, H., Herrera-Navarro, A.M., Morales-Hernández, L.A., Terol-Villalobos, I. R.: Morphological rational operator for contrast enhancement. J. Opt. Soc. Am. A **28**(3), 455–464 (2011)
5. Sui, X., Chen, Q., Bai, L.: Detection algorithm of targets for infrared search system based on area infrared focal plane array under complicated background. Optik-Int. J. Light Electron. Opt. **123**(3), 235–239 (2012)
6. Khan, J.F., Alam, M.S., Bhuiyan, S.: Automatic target detection in forward-looking infrared imagery via probabilistic neural networks. Appl. Opt. **48**(3), 464–476 (2009)
7. Shaik, J., Iftekharuddin, K.M.: Detection and tracking of targets in infrared images using Bayesian techniques. Opt. Laser Technol. **41**(6), 832–842 (2009)
8. Cao Yuan, Liu RuiMing, Yang Jie: Small target detection using two-dimensional least mean square (TDLMS) filter based on neighborhood analysis. Int. J. Infrared Millim. Waves **29**(2), 188–200 (2008)
9. Kim Sungho, Yang Yukyung, Lee Joohyoung, Park Yongchan: Small target detection utilizing robust methods of the human visual system for IRST. J. Infrared Millim. Terahertz Waves **30** (9), 994–1011 (2009)
10. Shao Xiaopeng, Fan Hua, Lu Guangxu, Xu Jun: An improved infrared dim and small target detection algorithm based on the contrast mechanism of human visual system. Infrared Phys. Technol. **55**(5), 403–408 (2012)
11. Deng He, Liu Jianguo, Chen Zhong: Infrared small target detection based on modified local entropy and EMD. Chin. Opt. Lett. **8**(1), 24–28 (2010)
12. Wang Xin, Liu Lei, Tang Zhenmin: Infrared dim target detection based on fractal dimension and third-order characterization. Chin. Opt. Lett. **7**(10), 931–933 (2009)
13. Deng He, Liu Jianguo: Infrared small target detection based on the self-information map. Infrared Phys. Technol. **54**(2), 100–107 (2011)
14. Seo, Hae Jong, Milanfar, P.: Static and space-time visual saliency detection by self-resemblance. J. Vis. **9**(12), 1–27 (2009)
15. Seo Hae Jong, Milanfar, P.: Visual saliency for automatic target detection, boundary detection, and image quality assessment. In: Proceedings of Acoustics Speech and Signal Processing (ICASSP), pp. 5578–5581. (2010)
16. Zhao Jufeng, Feng Huajun, Xu Zhihai, Li Qi, Peng Hai: Real-time automatic small target detection using saliency extraction and morphological theory. Opt. Laser Technol. **47**, 268–277 (2013)
17. OTCBVS Benchmark Dataset Collection. http://www.vcipl.okstate.edu/otcbvs/bench/

Participatory and Collaborative Design in the Creation of Applications for Societal Development

Bimlesh Wadhwa, Clarence Cai, and Colin Tan

Abstract This paper discusses the significance of participation with end users with the other stakeholders and multidisciplinary collaboration in the design process for the creation of applications for societal development. We have applied the participatory design approach in two different scenarios: (1) in the context of designing tablet apps for autistic children and (2) in the context of designing a chat room for math students. In each context, we (a) inform the design process through a background research and (b) incorporate participatory design and collaboration with stakeholders into the design and evaluation process. Presenting both cases, we intend to discuss the problems and opportunities of participatory approaches for designing applications for societal development. In conclusion, the participatory design process allows highly particular responses and reactions of stakeholders which enriches the evolutionary design process.

Keywords Participatory design • Tablet apps for autistic spectrum • Chat room design

---

B. Wadhwa (✉)
School of Computing, National University of Singapore, Singapore, Singapore
e-mail: dcsbw@nus.edu.sg

C. Cai
Global Business Services, IBM Singapore Pte Ltd., Singapore, Singapore
e-mail: clarence@sg.ibm.com

C. Tan
Group Digital Life, Singapore Telecommunications Limited, Singapore, Singapore
e-mail: wirelessed@gmail.com

E.B. Rajsingh et al. (eds.), *Informatics and Communication Technologies for Societal Development*, DOI 10.1007/978-81-322-1916-3_3, © Springer India 2015

1 Introduction

Originally, the design has been linked to the production of industrial systems. However, in the past few decades, design thinking has undergone many transformations, resulting in its scope extended from industrial systems to service economy to design for societal development. Victor Papanek wrote [1] that "design has become the most powerful tool with which man shapes his tools and environments (and, by extension, society and himself)." The idea of social design thinking is inspired by such ideas that designers have a responsibility and cause real change in the world through good design. The designer investigates the problem from a human-centered view, inquires its context, and systematically generates possible ideas to meet users' needs.

Participatory design [2] in interaction design is not just about "requirements for system developers and matching the look and feel with users' preferences," but it is about "an ethical argument" for giving the users an opinion in the design of the technology. A shared understanding of needs and wants is often the important ingredient for a successful working relationship. Also, the sooner it is introduced into the design process, the better. Here are some well-researched strategies [3]:

- Communicate early and often. Communicating and sharing plans and thinking early in the process can actually save a lot of time down the road. Sketching ideas together visualizes design concepts and builds an understanding that holds everyone together as one moves forward.
- Share the design process. Design is generally a chaotic process. Introducing it to users could be challenging, but it presents opportunities to engage users in new ways and should be embraced.
- Demonstrate the value of participation and collaboration throughout the process. Users' participation should not be limited to just the strategy.
- Develop trust and partnership. Understand that stakeholders rarely know what designers know – and vice versa.

In this paper, we present two different scenarios: (1) the context of designing tablet apps for children with autistic spectrum disorder (ASD) and (2) the context of designing a chat room for math students with an aim to discuss the significance of user participation and collaboration in the design process. In each context, we (a) inform the design process through a background research and (b) incorporate user participation into the design process. In the first case, we have developed two iPad apps where we worked with parents, teachers, and autistic children. In the second case, we developed a prototype for a chat room for students and teachers in a junior college. Through these experiences, we intend to discuss the problems and opportunities of participatory approaches for designing applications for societal development. The primary data in both scenarios were obtained by means of contextual inquiry, shared interpretation, site visits, interviews, and observations with participants in their respective context. The secondary data were collected by means of the evaluation of the available literature as well as documents that describe the users studied and successful cases of application of similar designs. In the sections below, each case is described considering two dimensions: (1) background research and (2) user participation and collaboration in the design process.

2 Tablet Apps for Children with ASD

Research in computer-based intervention for children with ASD has proven that children with ASD seem to be "drawn to technological devices," and it is important that researchers devise treatments that "take advantage of this fascination" [4]. In some cases, they even learn faster using computers than with traditional teaching strategies [5]. In this work, we use iPad to support quiz and storytelling activities and have developed two apps titled "I Know" and "I Can Tell." We believe that these apps, though developed with an aim to enhance language skills, could also enhance social behaviors such as collaboration, expression, and appreciation for pair activities [6].

2.1 Background Research to Inform the Design Process

2.1.1 Review of Apps Designed for Children with ASD

We evaluated existing apps TapSpeakChoice, iCommunicate, Touch Talking, Talking Cards, and iPrompts XL based on the following criteria:

- Usability: Match between system and the real world, consistency and standards, error prevention
- Is the UI easy to understand?: Visibility of system status, recognition rather than recall
- Esthetic and minimalist design
- Customizability: User control and freedom, flexibility and efficiency of use
- Feature set: What can the app do?

2.2 Design Considerations for Children with ASD

2.2.1 Simplicity and Practicality

The effect of overburden is amplified in children with ASD especially when children are overburdened with creative responsibilities. Aiming for "maximum involvement" can end up overburdening the children, which may "result in disengagement" [2]. Strong and sustained collaboration requires significant time, commitment, and resources. It is important to recognize that teachers, parents, and caretakers stand to gain very little by design participation and their otherwise demanding life routines get disrupted. Visits, interviews, and evaluations could be time-consuming on their part, and often, they don't see a tangible benefit until late in the development. Keay-Bright and Howarth [7] suggested that "whilst some learners with autism may enjoy a mode of interaction that indicates a right or wrong answer as it eliminates confusion, for less able children such indications of failure can cause extreme anxiety and meltdowns, which are barriers to learning and sociability." Simplicity is therefore important, but it should not be too black-and-white, which becomes somewhat a barrier to learning as less able children may be discouraged by the wrong answers.

2.2.2 Story Paradigm

Gal et al. [8] used a Story-Table, aimed at HFA individuals, that employs multi-touch technology to allow up to four users to interact with it at the same time. This system affords the users the opportunity to tell stories between each of them. The study has reported "increased ability to communicate verbally" as well as the decrease in maladapted mannerisms (e.g., hand flapping) while using the Story-Table.

2.2.3 Quiz Paradigm

A typical way to test the knowledge of an individual with ASD is by quizzing him/her. The quiz paradigm is simple and easy to administer, and the results are easy to retrieve and store as well. Silver and Oakes [9] use quiz paradigm to teach people with autistic spectrum disorders to better recognize and predict emotional responses in others.

2.2.4 Self-Fading Prompts

Prompt fading is used to assist the child or adult with ASD to get used to the task at hand without pointing out the correct answer. "Prompts are useful in helping people display new, desirable behaviour, but new skills are mastered . . . only if prompts can be removed" [10]. In the decreasing assistance prompting technique, the "learners receive what assistance they need to perform a new skill when instruction begins." Stimulus fading is a prompting technique that "exaggerates some physical dimension" such as color, size, and intensity. This includes not only visible prompts but also auditory and all other senses. Then, gradually, this prompt is reduced until it is either gone or some aspect of it is left behind.

2.3 *Stakeholder Participation and Collaboration in the Design Process*

We made several observations of autistic children by visiting each of the special schools St. Andrew's Autism Centre, Lighthouse School, Rainbow Centre, and Eden School at Singapore. The participatory design of the apps was conducted with four users: parent of a high-function autistic child, parent of an autistic child, grandparent of a typically developing child of the same age group as the autistic children under study, and a teacher teaching autistic children. Users were briefed about the research objectives and their role in participatory design and evaluation. Observations, structured and semi-structured interviews, and open-ended questionnaires were used to gather data and feedback on paper and interactive prototypes developed as part of the iterations. Shared insights shaped the personas, the scenarios, and the design of the apps. The design goals of the apps were set as simple, predictable, visual, mistake-free, and open-ended and take advantage of children's strengths. Iterative development continued for a period of four months for the two apps (see Tables 1 and 2).

3 Collaborative Discussion in a Multitasking Environment

The method of text-based chat has remained similar throughout the years, with single-line text messages being sent to and fro. In the classroom learning environment, real-time chat systems are "quasi-synchronous" – while the participants in a chat room are interacting simultaneously, their messages are read by the other parties only after they are processed and sent through the network [11]. Most of the implementations, such as the Integrated Virtual Learning Environment (IVLE) Chat in the National University of Singapore (NUS), are too simple and inefficient. Moreover, when filled with too many users chatting simultaneously, such chat systems end up becoming complicated, chaotic, and hard to keep up. They do not leverage on the strengths of discussion in a traditional classroom setting, such as a common collaborative space in the form of a whiteboard, and turn taking regulated by an educator or teacher. Users also tend to multitask while chatting and hence not pay full attention to the chat. Hence, chat systems need to acknowledge the presence and absence of such user activity. Moreover, math symbols are a need for online discussions and lacking in most chat systems. Besides, they must be free of complex notations such as LaTeX in order to be easy to use for middle and high school students. Our goal therefore was to attempt to design and implement a user interface prototype for a collaborative problem-solving math chat, incorporating mathematical symbols.

Table 1 Summary of quiz app

App description	Features
App title: I Know A quiz app. It enables children to identify a specific object. The features allow the parent or caretaker to create/edit/delete the quiz. The quiz can be based on the photos of the objects taken from the child's environment. An audio clip for each object can be recorded in the familiar voice of the parent or caretaker. The child is given a feedback on his or her action. Negative feedback is kept to a subtle level as compared to the positive feedback	Features Quiz management Create/edit/delete Share quiz Quizzing features Play quiz Display quiz results

Table 2 Summary of storytelling app

App description	Features
App title: I Can Tell A storytelling app. It enables children to build a story from everyday objects and situations. It allows a collaborative story building with the parent or caretaker. The features allow the parent or caretaker to take photos of the objects from the child's environment or even downloaded from the Internet. Pictures can be organized in specific folders and tapped to bring to the story area of the screen. A picture can be moved, resized, or rotated to provide interaction and to make interesting stories. Sidebars are provided on both sides of the screen so that the parent or caretaker and the child can both utilize the screen at the same time. This allows the app to augment typical noncomputer storytelling and add a multimedia multi-touch element to it. In addition, all actions on the screen in building the story can be recorded. This recording is done with the audio of what is spoken recorded as well. Recordings can be saved and played back as and when needed. A playback of the story can be very useful in helping the child to develop language skills. Playbacks also allow the child to repeatedly view previously created stories so that the child can have the benefit of having the story told multiple times without having the parent/caretaker to retell it again and again	Features Storytelling features Tap pictures Move/resize/rotate pictures Input text Recording/playback Record storytelling Able to record all actions taken on screen Playback recordings Pictures management Add/delete pictures into app Organize pictures in folders

3.1 Background Research to Inform the Design Process

3.1.1 Quasi-synchronous Chat

Past research has surfaced many problems with quasi-synchronous chat. We specifically studied [12] problems of phantom responsiveness, phantom adjacency pairs and turn taking, virtual simultaneity, silence, pauses and visibility, co-text loss, message identity, useful history and persistence, flooding, and repair and reviewed the following.

3.1.2 Guidelines and Frameworks

There are a number of frameworks which have been developed in the past. A few, The Disco Framework [13], Change Awareness [14], and Awareness Support [15], were revisited for the design purpose here.

3.1.3 The Math Problem

A math-enabled learning management system [16] is expected to include integrated equation tool, sketching and drawing of diagrams, highlighting, pointing or referencing, shared document editable in real time, sound to speak out equations, ability to view equations as they are being typed and built upon, reusability of work to be copied/pasted, and gestures or use of avatars. Mühlpfordt and Wessner [17] suggest "dual-interaction spaces," while Stahl [18] suggests integrating chat rooms with "asynchronous and synchronous media" such as wikis and "deictic referencing" using lines or language to link to another drawing or message.

3.1.4 Chat Applications from Research Labs

We reviewed research-based applications: Chat Circles [19] by MIT Media Lab; Loops and Babble [20] by IBM T. J. Watson Research Center; Threaded Chat [21], Comic Chat [22], and Flow Client [23] by Microsoft Research; and ConcertChat [17] by the Fraunhofer Institute for Integrated Publication and Information Systems.

3.1.5 Commercial Chat Applications

We reviewed a few text-based chat modules of commercially available learning management systems (LMS), e.g., Blackboard, Moodle, Second Life, and Google Wave.

3.1.6 Existing Math Chat Tools

WebEQ, NetTutor, Windows Live Messenger, enVision platform for its dual-space model, MathIM, and IMathAS are a few chat tools reviewed for their most accepted features by the users.

Table 3 summarizes [12] the chat applications mentioned above.

Table 3 Summary of chat applications

Chat application	Model	Collaboration tools	Main problems addressed (section 0)
Chat Circles	Graphical	Chat only chat only	Awareness, useful history
Loops and Babble	Sequential	Chat, bulletin board, tabs	Awareness, persistence, useful history
Threaded Chat	Threaded	Chat only	Adjacency, simultaneity, repair
ConcertChat	Sequential	Chat, whiteboard	Co-text loss, repair, referencing
Comic Chat	Graphical	Chat only	Useful history, repair
Flow Client	Graphical, sequential	Chat only	Adjacency, links between people and what they say, repair
IVLE Chat	Sequential	Chat only	N/A
Blackboard	Sequential	Chat, whiteboard, browser	Adjacency, flooding
Moodle Chat	Sequential	Chat only	N/A
Second Life	Graphical	Chat, whiteboard	Turn taking
Google Wave	Threaded	Chat, whiteboard, document editing, wiki	Adjacency, simultaneity, repair, turn taking, flooding, useful history
enVision	Sequential	Chat, whiteboard	Referencing
MathIM	Sequential	Chat only	N/A
IMathas	Sequential	Chat only	N/A

3.2 *Stakeholder Participation and Collaboration in the Design Process*

Armed with background research, we conducted a contextual inquiry with stakeholders, students, and teachers to propose a user interface DiscussionZone – for a classroom e-learning environment. With the context of a real client, a junior college, we approached the development of a math chat interface by adapting parts of the Rapid Contextual Design (CD) methodology [24]. A series of site visits, interviews, discussions, observations, and participation resulted in a set of requirements, notably simple user interface, insertion of math symbols or drag-and-drop from an icon library based with constraints on choices, no complex LaTeX commands, a "dual-space" model, and restricted instructors' controls. Stakeholders shared their experiences with chat systems, notably server slowness and missing element for evidence of learning. A shared interpretation resulted in a wish list categorized in UI, chat, collaborative space, history and log, efficiency and reliability, and shared design goals: (1) The interface should mimic classroom discussion with the dual-space model but it should not be too unfamiliar for regular chat users, and (2) students should be able to chat, contribute, and learn easily, while educators should be able to observe and extract the students' learning process. Based on the design goals and wish list, sketches and storyboards were constructed before arriving at the final design for a prototype. The overall attempt was to bridge

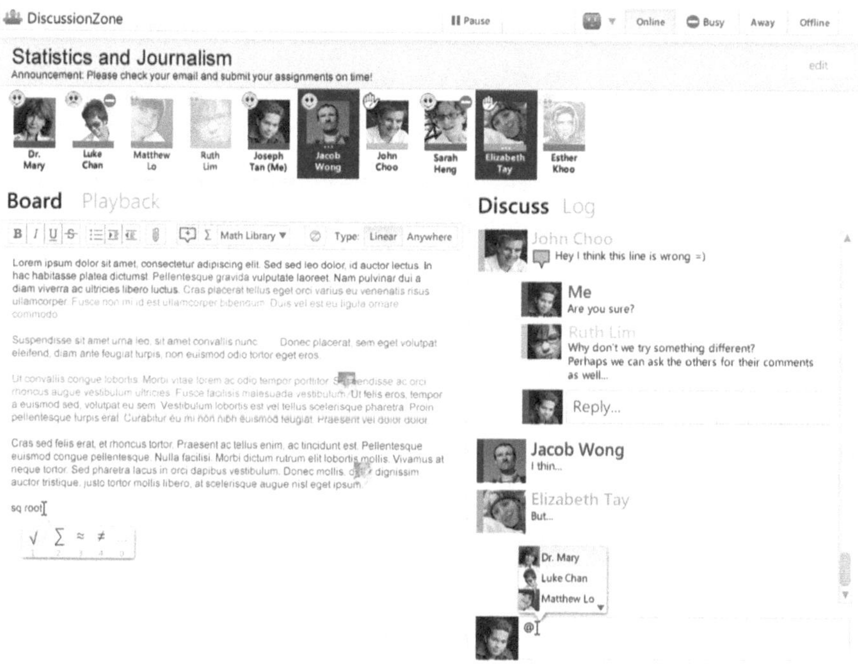

Fig. 1 Sample interface

the gap between face-to-face communication and online chat, addressing the issues identified in background research. We found the following solutions:

- *Phantom responsiveness, adjacency, virtual simultaneity, flooding*: Display messages as they are typed in real time, character by character, keystroke by keystroke
- *Silence, pauses, and visibility*: Having intelligent status awareness such as auto status detection and use of emoticons
- **Co-text** *loss*: Referencing and in-line comments
- **Message** *identity*: Using profile pictures and color coding
- *Useful history*: Playback of discussion and chat log
- *Repair*: Facilitating repair by improving @ mentions, private messaging
- *Math symbols*: By using auto-prediction (Fig. 1)

The design of each element – user list, topic, board, in-line comments, math symbols, discussion log, @ mentions, private messaging, status update, resuming from non-online status, persistent emotions, playback, moderator tools, activity graph – was rationalized and shared with the stakeholders through a series of evaluations using usability evaluation of task-based scenario.

4 Conclusion

In this paper, we sought out to establish how participation design and collaboration play a significant role in designing for societal development. We have presented two such designs – tablet apps in assisting or teaching children with ASD and a chat room design to facilitate collaborative discussion in a multitasking environment.

Participatory design is an essential and beneficial technique for developing a usable tablet app. It allows us to view through a window and into the lives of these children and their caretakers. A thorough user study and continuous user evaluation can help develop a usable app for the child with ASD as well as for his or her caretakers. In order to understand the overlaps between the goals of "Design for All" and the opportunities afforded by specialized technologies and to develop a tablet app that is beneficial to a typically developing child as well, we included the caretakers (parents and teachers) in the design process, starting with paper prototypes, then moving on to actual prototypes, and finally, usability testing of the final product. A bonus validation of the product was that the "I Know" and "I Can Tell" apps are both approved by Apple.

The user interface of a chat room is designed with an intention to solve the problems of quasi-synchronous communication reviewed in the past research, as well as to propose a better interface for online classroom discussion. With the stakeholders' participation, we discovered that the evidence of learning is very important after an online discussion. The evidence includes the following: How much did a student learn? Who is leading the discussion the most? Who knows their topics well? Natural language processing, sociology, conversational analysis, and other fields of knowledge are required for an effective design here.

In conclusion, the participatory design methods used here represents a synthesis of methodologies undertaken by interaction designers which involve the participants in the design process. We extended these methods by inviting users to contribute to the design of the software during critical stages of development. Although the methods were time-consuming, partly because the problems are so complex but also the need to become accepted by the participants, we ensured that any proposed solutions were designed from the point of view of the user rather than the technology. Participatory design requires that people, whose lives stand to be affected in some way by the outcomes of the research, are given the opportunity to identify themselves within the objectives of the research. We believe that the way in which users, e.g., autistic children in one of the scenarios here, interact cannot be presented through statistical data or empirical studies; therefore, gaining understanding through their actions, body activities, expression, and language becomes paramount to any application development for their benefit. User participation and collaboration in the design process presents significant challenges, for example, many autistic children do not use verbal language to communicate and find the slightest change in their routines extremely painful. Becoming accepted by them was crucial and so the observations had to be very carefully carried out. A critical research phase was undertaken before approaching collaborative partners in order to develop an informed understanding of the issues. An iterative cycle, research, inquiry with observation and shared interpretation, and development, was followed throughout the development of two designs. We noted that in the collaborative design for societal development, an attitude from designing for users to one of designing with users is more beneficial. Participatory design is the belief that users offer crucial insights to the design process When all three perspectives (what people do, what they say, and what they make) are explored simultaneously, one can more readily understand and establish empathy with the users who use such interactive products. The designs can, in fact, harness the collective and infinitely expanding set of ideas and opportunities that emerge when the users who have a stake in the process are invited to "be a part."

References

1. Papanek, V.: Design for the real world human ecology and social change. Thames and Hudson, London (1985)
2. Porayska-Pomsta, K.: Developing technology for autism: an interdisciplinary approach. Pers. Ubiquit. Comput. **16**(2), 117–127 (2012)
3. Polgar, J.M.: The myth of neutral technology: design and use of assistive technology, pp. 17–23. Springer, New York (2010)
4. Colby, K.M.: The rationale for computer-based treatment of language difficulties in nonspeaking autistic children. J. Autism Child. Schiz. 1–13 (1973)
5. Bosseler, A., Massaro, D.W.: Development and evaluation of a computer-animated tutor for vocabulary and language learning in children with autism. J. Autism Dev. Disord. **33**(6), 653–672 (2003)
6. Wadhwa, B., Cai, C.: 11th Asia Pacific conference on computer human interaction, APCHI 2013, Bangalore, India, 24–27 Sept 2013
7. Keay-Bright, W., Howarth, I.: Is simplicity the key to engagement for children on the autism spectrum? Pers. Ubiquit. Comput. **16**(2), 129–141 (2012)
8. Gal, E.: Enhancing social communication through story-telling among high-functioning children with autism. In: Proceedings of the first international conference on intelligent technologies for interactive entertainment (INTETAIN'05), pp. 320–323. (2005)
9. Silver, M., Oakes, P.: Evaluation of a new computer intervention to teach people with autism or asperger syndrome to recognize and predict emotions in others. Autism 5(3), 299–316 (2001)
10. MacDuff, G.S., Krantz, P.J., McClannahan, L.E.: Prompts and prompt-fading strategies for people with autism. In: Maurice, C., Green, G., Foxx, R. (eds.) Making a difference: behavioral intervention for autism, pp. 37–50. Pro-Ed, Austin (2001)
11. Garcia, A., Jacobs, J.B.: The eyes of the beholder -understanding the turn taking system in quasi-synchronous computer mediated communication. Res. Lang. Soc. Interact. **32**(4), 337–367 (1999)
12. Wadhwa, B., Tan, C.: DiscussionZone – collaborative discussion in a multi-tasking environment. Presented at 4th international conference on human computer interaction, Pune, India, 18–21st Apr 2012
13. Holmer, T., Lukosch, S., Kunz, V.: Addressing co-text loss with multiple visualizations for chat messages. Groupware: design, implementation, and use. Springer, Berlin, Heidelberg, City (2008)

14. Gutwin, C., Graham, T.C.N., Wolfe, C., Wong, N. Alwis, B.D.: Gone but not forgotten: designing for disconnection in synchronous groupware. In: Proceedings of the 2010 ACM conference on Computer supported cooperative work, Savannah, Georgia, USA (2010)
15. Tam, J., Greenberg, S.: A framework for asynchronous change awareness in collaborative documents and workspaces. Theor Empirical Adv Groupware Res **64**(7), 583–598 (2006)
16. Leventhall, L.: Bridging the gap between face to face and online maths tutoring. In: Proceedings of the international congress of mathematics education-10, Copenhagen, Denmark (2004)
17. Mühlpfordt, M., Wessner, M.: Explicit referencing in chat supports collaborative learning. In: Proceedings of the 2005 conference on Computer support for collaborative learning: learning 2005: the next 10 years! Taipei, Taiwan (2005)
18. Stahl, G., Zemel, A., Sarmiento, J., Cakir, M., Weimar, S., Wessner, M., Muhlpfordt, M.: Shared referencing of mathematical objects in online chat. International Society of the Learning Sciences, City (2006)
19. Viegas, F.B., Donath, J.S.: Chat circles. In: Proceedings of the SIGCHI conference on human factors in computing systems: the CHI is the limit, Pittsburgh, PA, USA (1999)
20. Erickson, T., Kellogg, W.A., Laff, M., Sussman, J., Wolf, T.V., Halverson, C.A., Edwards, D.: A persistent chat space for work groups: the design, evaluation and deployment of loops. In: Proceedings of the 6th conference on designing interactive systems, University Park, PA, USA (2006)
21. Erickson, T., Smith, D.N., Kellogg, W.A., Laff, M., Richards, J.T., Bradner, E.: Socially translucent systems: social proxies, persistent conversation, and the design of Babble. In: Proceedings of the SIGCHI conference on human factors in computing systems: the CHI is the limit, Pittsburgh, PA, USA (1999)
22. Kurlander, D., Skelly, T., Salesin, D., Comic Chat.: In: Proceedings of the 23rd annual conference on Computer graphics and interactive techniques (1996)
23. Vronay, D., Smith, M., Drucker, S.: Alternative interfaces for chat. In: Proceedings of the 12th annual ACM symposium on User interface software and technology, Asheville, NC, USA (1999)
24. Beyer, H., Holtzblatt, K., Baker, L.: An agile customer-centered method: rapid contextual design extreme programming and agile methods – XP/Agile Universe, pp. 50–59. (2004)

Cloud Service Brokers: Addressing the New Cloud Phenomenon

Bimlesh Wadhwa, Aditi Jaitly, Nidhi Hasija, and Bharti Suri

Abstract The concept of cloud computing with on-demand provisioning of a shared pool of configurable computing resources of unprecedented size, elastic scalability, and ubiquitous network service accessibility has gained huge momentum in the IT industry and is high on the agenda of the industry and research community from developer as well as from end-user perspective. However, there are many issues faced in migrating existing systems into the cloud and between providers that have led to its slow adoption. Cloud service brokers (CSBs) have emerged as a solution to address this growing need and provide services that facilitate smooth cloud adoption, management, migration, and maintenance for cloud users and providers. This chapter addresses this new cloud phenomenon by reflecting on cloud adoption and migration challenges and presenting an overview of CSB utilities and the proposed concept of CSB registry.

Keywords Cloud service broker • Cloud migration • Cloud adoption

B. Wadhwa (✉)
School of Computing, National University of Singapore, Singapore, Singapore
e-mail: dcsbw@nus.edu.sg

A. Jaitly
Faculty of Management Studies, University of Delhi, New Delhi, India
e-mail: aditi.j15@fms.edu

N. Hasija
Tata Consultancy Services Ltd., New Delhi, India
e-mail: nidhi.hasija@tcs.com

B. Suri
University School of ICT, GGS Indraprastha University, New Delhi, India
e-mail: bhartisuri@gmail.com

E.B. Rajsingh et al. (eds.), *Informatics and Communication Technologies for Societal Development*, DOI 10.1007/978-81-322-1916-3_4, © Springer India 2015

1 Introduction

Cloud is believed to be the primary driving force for IT innovation. Organizations are increasingly gaining a competitive advantage owing to the profit and revenue benefits that cloud delivers. IBM's survey of leading businesses revealed that the strategic importance of cloud to decision makers of organizations is bound to double in the near future [1]. However, cloud poses a fair set of technical and managerial challenges, pertaining to cloud adoption and maintenance, one of them being the highly fragmented cloud space with multiple providers. Consequently, these limitations and challenges have turned out to be precursors of a novel concept of cloud service brokerage. Cloud service brokers (CSBs) have emerged as a solution to address the need of consultancy, management, and technology for both the cloud users and providers [2]. They promise to help cloud providers in the entire life cycle from cloud adoption to management and maintenance.

According to the research firm Gartner, the CSB sector currently represents one of the fastest growing sectors of the cloud and is expected to grow to a whopping $100 billion in revenue by the end of 2014 [3]. Currently, there are several CSBs in a highly fragmented market, each providing both distinct and overlapping set of services [4]. This makes it increasingly difficult for consumers to select a CSB. We therefore introduce the concept of a dynamic CSB registry that helps facilitate the selection of a CSB. This chapter, in the sections below, (1) summarizes cloud adoption and migration challenges, (2) presents a consolidated view of the CSB utilities, and (3) introduces the concept of CSB registry.

2 Cloud Adoption and Migration

2.1 Cloud Adoption

Even though the cloud offers numerous benefits including strategic flexibility, cost-effectiveness, scalability, and ease of operation, there are also many issues and risks involved which have led to slow cloud adoption. Cloud adoption here is considered as moving the existing data or applications to the cloud. Some of the challenges faced by the cloud users are as follows.

2.1.1 Cost-Benefit Analysis

Existence of numerous cloud offerings loaded with a myriad of features makes the process of choosing and adopting the cloud a challenge. Choosing the optimal configuration with minimum cost is often an essential requirement for cloud adoption, especially for small and medium enterprises [5]. Usually the cost analysis includes usage costs, expansion costs, cost of lowered QoS, availability, and delays

[6, 7]. Costs associated with lapse of service-level agreements (SLAs), contract termination, and data migration are additional costs to be considered in cost-benefit analysis.

2.1.2 Security Concerns

Since a company's sensitive information is surrendered to a third-party cloud provider, the company potentially comes under great risk. It also gets prone to hack attacks and threats. However, this problem can be addressed to an extent by including security clauses in SLAs.

2.1.3 Integration of Services Within or Across Cloud Offerings

As consumers adopt more and more cloud applications, integrating SaaS (software as a service) applications with existing on premise applications or other cloud applications becomes a huge concern [8].

2.2 *Cloud Migration*

With rapidly increasing popularity of cloud, cloud users and providers are facing the issue of cloud migration. Due to a dynamic business environment, the users may need to periodically review their strategy their cloud strategy in order to factor in scenarios like business expansion, issues with existing cloud provider, etc. Cloud migration is largely unexplored, and deep diving into it exposes many technical and managerial challenges [9], as depicted in Table 1.

3 Cloud Service Brokers: A Solution

A CSB not only assists in cloud adoption and management but also helps in cost saving and in providing faster agility along with more flexible IT capability [10]. Figure 1 shows it as an intermediary between consumers (C1, C2,. . .Cn) and providers illustrating interaction between the consumer and the CSB, the CSB and the cloud provider, and the consumer and the cloud provider. The direct shell interaction between consumer and provider is further elaborated in this section (see U14).

In this section, we have enumerated services that a CSB can provide and classified them based on different parameters that are as follows:

- Business service versus technical service
- Consumer-specific versus provider-specific utilities

Table 1 Snapshot of cloud migration challenges

Technical challenges	*Description*
Vendor lock-in and dependencies	High exit costs owing to: Different file formats Paucity of migration tools Lack of standard cloud solutions
Need for data backup	Possibility of data loss during data migration
Impediment to business continuity	Need for a phased migration Higher costs
Managerial challenges	*Description*
SLAs laws and agreements	Measuring and monitoring cloud performance is complex Need to evaluate SLAs
Cost analysis	Detailed cost analysis needed to evaluate profitability
Evaluating CSP credentials	Need to choose a trusted CSP
Privacy laws and location	Understand and manage intercountry privacy laws as CSP servers are distributed across the globe

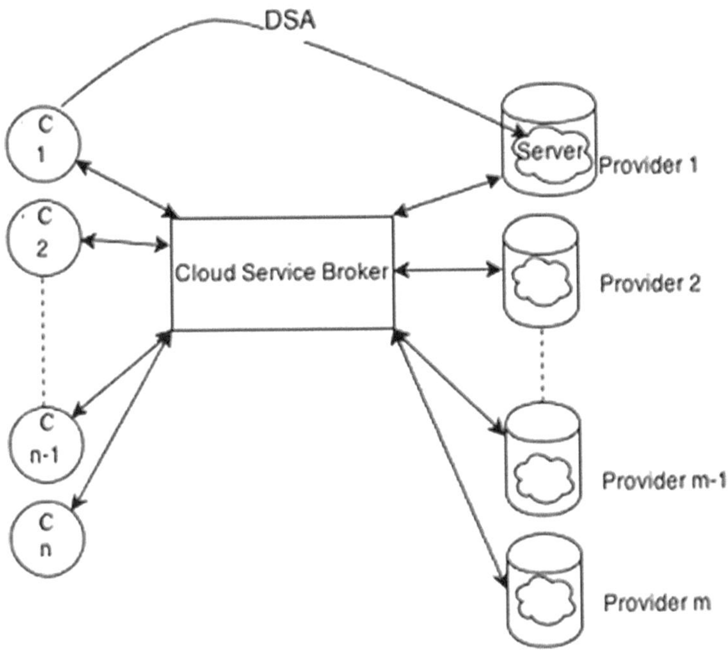

Fig. 1 Interaction scenario

There are four distinct roles of cloud service brokers, as mentioned below [11]. Gartner suggests that by 2014, IT organizations in 30 % of the 1,000 global companies will broker (aggregate, integrate, and customize) two or more cloud services for internal and external users, up from 5 % today [12].

3.1 Aggregator (R1)

It is a broker that takes multiple cloud services encompassing different providers and funnels them to clients through a channel. The client uses the services proportionate to one's needs and pays one's bill to the broker. The broker "owns" the relationship on behalf of the client. The broker might manage service-level agreements (SLAs) for the customer. It can offer extra services on top of the cloud, ranging from extra security services to business continuity and anything in between.

3.2 Integrator (R2)

Technically skilled, this type of provider merges services from an assortment of cloud providers to create a net new business process.

3.3 Governance Services (R3)

This is especially important in highly regulated industries such as healthcare and financial services, where protection of consumer data is essential. Here, the broker ensures that the cloud service provider treats the data in accordance with the regulations and policies and that security practices are adhered to. It will handle the basic administrative tasks of the system.

3.4 Customizer (R4)

As the name suggests, customizers will create extensions to existing cloud services for a specific functionality and then ensure that the extensions work properly. It will alter its functionality to the needs of the consumer.

3.5 Overview of CSB Services

In this subsection, we provide a consolidated list of CSB services (U1–U18) [13].

3.5.1 U1: Facilitating Cloud Adoption

Cloud adoption can be termed as moving the existing data or applications to the cloud. A CSB should provide cloud engineering guidance by helping weigh the advantages and drawbacks of cloud [14, 15]. It should also provide a commercial viability check followed by choice of, engagement with, and management of cloud providers.

3.5.2 U2: Scalability

The CSB should ensure that the consumer is provided cloud resources in proportion to its business needs by negotiating with the cloud providers.

3.5.3 U3: Solutions Provided and Implementation Services

A CSB should provide software or products to automate the processes of cloud adoption, cloud-sourced development, cloud metrics estimation, data security, and cloud integration [8, 16].

3.5.4 U4: Customer Support and Guidance

A CSB should provide continuous support and guidance to the consumers, help them strategize and plan, and also suggest better alternatives both in terms of usage and providers selection [16]. Also, it should mediate for complaints and grievances redressal and consequently ensure customer satisfaction.

3.5.5 U5: Updating with Time and Business Change

A CSB should help alter the cloud strategy of the consumer and aid consumers in handling changes of business processes [17].

3.5.6 U6: Integration Services

A CSB should help identify the need and strategize the integration process. It can provide automated tools to aid, manage, and maintain integrations [16, 18].

3.5.7 U7: Handle Enhancements in the Cloud

The version of the cloud might change over time, especially if it is a public cloud. The CSB should ensure the compatibility of consumer's applications to the enhanced versions of the cloud [8].

3.5.8 U8: System Administration

System administration entails tasks like defining monitoring settings, compliance standards, and cloud policies [8, 19]. It also involves features such as writing and maintaining the SLA (service-level agreement), extending the pricing model, creating data backups, and ensuring cloud recovery. The administration of the system should be a responsibility of the CSB.

3.5.9 U9: Data Security

Data security includes the most essential prerequisite of data migration onto the cloud. CSB must ensure the creation of a well-drafted SLA with clearly defined security clauses [20].

3.5.10 U10: Customization

Customization involves tailoring the services in accordance to the need. A CSB should help in need-based tailoring of services and thereby saving costs [21].

3.5.11 U11: Aggregation and Multiple Cloud Provider Support

A CSB should help aggregate cloud services from different cloud providers (private or public cloud) into one or more services by means of a transparent process [20, 22].

3.5.12 U12: Interoperability Between Providers and Aggregation

A CSB should ensure interoperability between multiple clouds and providers it supports [23]. It should ascertain that a consumer is able to deploy its interconnected applications with multiple providers at the same time.

3.5.13 U13: Cloud Monitoring

Cloud monitoring includes activities such as capacity and resource planning, capacity and resource management, data center management, billing, SLA management, troubleshooting, performance management, and security management [24]. For the cloud providers, its output includes workload information, while for the consumer it implies information bearing QoS (quality of service). A CSB should monitor the cloud servers and give the consumer a summary of CPU, memory, storage, and network usage in clearly arranged diagrams.

3.5.14 U14: Direct Shell Access to the Consumer

A CSB should deliver the consumer direct shell access to the servers, so that it can handle its own administrative tasks when required from their Web browsers [25].

3.5.15 U15: Avoiding Vendor Lock-In

The rise in number of cloud providers, their heterogeneous interfaces, and absence of interoperability may bind a consumer to a specific cloud provider thereby inhibiting switching [26]. A CSB should help preclude this process of vendor lock-in by including standardization clauses in the SLA fine print.

3.5.16 U16: Costs-Usage Analysis

Costs are a prime consideration of consumers that adopt cloud. A CSB should provide a cloud-cost calculator that helps consumers in knowing the cost-usage snapshot and make better decisions [27].

3.5.17 U17: Continuous Improvement in the Cloud

A CSB should also ascertain that the cloud providers recurrently update their services, technology, and infrastructure [8].

3.5.18 U18: Migration Brokerage

A CSB should help in the process of migration brokerage and help migrate consumer applications and data from one cloud provider to another, which is better matched to the needs of the consumer [25, 28].

Also each of these utilities can be classified based on the above roles and also business and/or technical utility [13].

Let Set A = {Un|1 < =n < =18}.
Set A includes consumer-specific utilities that help make the task of the consumer easier.
Set B = {U1, U3, U4, U8, U9, U13, U16}.
Set B includes provider-specific utilities.

4 Cloud Service Registry

Buyya et al. [29] talk about the concept of market registry to facilitate publishing and discovery of cloud service providers and services. Dawson et al. elaborate on the concept of a dynamic cloud service registry, which is regularly updated with the latest available abstracted cloud services [30]. Buyya et al. [29] bring out the concept of a cloud information service registry that performs the matchmaking role of mapping consumer requests to apposite cloud provider [31]. At the level of industry, the Cloud Security Alliance launched an initiative, namely, Security, Trust and Assurance Registry, to enable cloud consumers compare and review security practices of various cloud providers [32]. Summarily, the above initiatives taken up in the past have increased transparency levels, made the market more competitive, and have ultimately led to a better procurement experience.

However, we realize that there is a palpable gap with respect to a similar registry for cloud service brokers. Although the cloud brokerage space is in its nascence, it is highly fragmented with many players [4]. Daryl Plummer [33] harps on the importance of CSBs for companies that want to maximize their results. We gather that the significance of CSBs will increase in future. Therefore, in this section we propose the concept of a CSB registry. Research can help the industry with a novel concept of CSB registry whose features are as follows:

- Service Comparisons: The services provided by different CSB can be compiled on a single site. These can be coupled with information of the price charged for each of these services. This not only will serve the purpose of convenience in information retrieval but also will aid consumers in making better comparisons between a pool of CSB. This information will also be dynamically updated and kept up to date.
- Ratings: The existing consumers can rate and review their experiences with the CSBs. These evaluations can be through a questionnaire or descriptive reviews. The questionnaire can bear questions on their experience with different services. These can then be used to rate not just the CSBs on the whole but also specific services they offer.
- Certifications: Various CSBs can be offered certifications based on their service sets, customer satisfaction levels, initiatives taken, innovation levels, and other practices followed. These certifications will help CSBs in building consumer's trust. It will also serve the purpose of a rating parameter for consumers.

- A Platform to Market and Advertise: The CSBs can use this platform to differentiate their own set of services, market their core service features, and also increase their visibility in the consumers' mind. Some of the key advantages of the registry can be as follows:
  - Improved transparency and assurance levels
  - Ease of CSB selection
  - Creation of a competitive environment
  - Higher CSB adoption rate by consumers
  - CSB adherence to best practices
  - Mutual benefits for all players (consumers, CSBs, providers)

  The ease and convenience offered by it will help make the consumers view CSBs as not just an alternative but a prerequisite for cloud adoption or management.

5 Conclusion and Future Direction

In this chapter, we have summarized the adoption and migration issues that cloud users face and how CSBs address those issues by their services. We then proposed the concept of an industry implementable dynamic CSB registry that facilitates selection of a CSB. Summarily, we can say that cloud service brokers (CSBs) have emerged in the cloud landscape as a facilitator of cloud adoption, management, migration, and maintenance [13].

However, there is a lot to be done in this realm of CSBs. The academia and industry should strive to coproduce knowledge and dialogic relationships in this evolving area of CSBs [34]. Specific research goals should be formulated to synthesize and maximize rigor and relevance to the cloud providers and users. Generalized revenue and pricing models that can be implemented by the CSBs in the industry are becoming increasingly relevant. In the short term, landscape evaluation of CSB sector, research on additional CSB services, improving CSB architecture, and mathematical modeling of revenue streams could be useful areas to explore, while in the long term, research will need to come up with industry implementable revenue models, CSB registry along with improvements in services and security models. As has been rightly put, the journey of a thousand miles begins with a small step. This is perhaps the first call to the academia and industry to help augment the phenomenon of CSBs.

References

1. IBM: IBM Study. http://www-03.ibm.com/press/us/en/pressrelease/42304.wss
2. KPMG: The Cloud changing the business ecosystem. http://www.kpmg.com (2013)
3. Bova, T., Plummer, D.C., Petri, G., Malinverno, P., Govekar, M.: Cool Vendors in Cloud Services Brokerage, Gartner (2013)

4. Cloud Services Brokerage (CSB) Market Worldwide Forecasts and Analysis (2013–2018), marketsandmarkets (2013)
5. AkashRanka: Cloud Migration in start-ups, NUS (2013)
6. Khajeh-Hosseini, A., Greenwood, D., Sommerville, I.: Cloud migration: a case study of migrating an enterprise IT system to IaaS. In: Cloud Computing (CLOUD), IEEE 3rd International Conference on, pp. 450–457. IEEE (2010)
7. Kondo, D., Javadi, B., Malecot, P., Cappello, F., Anderson, D.P.: Cost-benefit analysis of cloud computing versus desktop grids. In: Parallel and Distributed Processing, IPDPS IEEE International Symposium on, pp. 1–12. IEEE (2009)
8. Appirio: Services. http://www.appirio.com/services/
9. García-Galán, J., Rana, O., Trinidad, P., Ruiz-Cortés, A.: Migrating to the Cloud: a Software Product Line based analysis (2011)
10. Lawler, C.M.: Cloud Service Broker, Hitachi Green IT Cloud Summit. http://www.hitachiconsulting.com/files/pdfRepository/Cloud-Service-Broker-Presentation-Green-IT-Cloud-Summit-2012.pdf
11. Paul, L.G.: Brokering the cloud: a guide to cloud service brokerage ChannelproSMB. http://www.channelprosmb.com/article/brokering-the-cloud-a-guide-to-cloud-service-brokerage
12. Cloud Computing: Gartner, http://www.gartner.com/technology/topics/cloud-computing.jsp
13. Wadhwa, B., Jaitley, A., Suri, B.: Cloud Service Brokers: an emerging trend in cloud adoption and migration. In: APSEC (2013)
14. Khajeh-Hosseini, A., Greenwood, D., Smith, J.W., Sommerville, I.: The cloud adoption toolkit: supporting cloud adoption decisions in the enterprise Software. Pract. Exp. **42**(4), 447–465 (2012)
15. Fellows, W.: Partly Cloudy, Blue-Sky Thinking About Cloud Computing, Whitepaper. 451 Group (2008)
16. Cloud Sherpas, http://www.cloudsherpas.com
17. Grivas, S.G., Uttam Kumar, T., Wache, H.: Cloud broker: bringing intelligence into the cloud in Cloud Computing. In: IEEE 3rd International Conference, IEEE (2010)
18. Integrating the Cloud: Bridges, Brokers & Gateways. White paper, F5 Networks, Inc. (2012)
19. Oracle Enterprise Manager: Cloud Administrative Guide. 12.1.0.2, Oracle (2013)
20. Nair, S.K., Porwal, S., Dimitrakos, T., Ferrer, A.J., Tordsson, T., Sharif, T., Sheridan, C., Rajarajan, M., Khan, A.U.: Towards secure cloud bursting, brokerage and aggregation. In Web Services (ECOWS), IEEE 8th European Conference, IEEE, pp. 189–196 (2010)
21. Samba, A.: Logical data models for cloud computing architectures. IT Professional 14.1, pp. 19–26 (2012)
22. Multicloud Platform: Right Scale. http://www.rightscale.com/products/multicloud-platform.php (2013)
23. Parameswaran, A.V., Chaddha, A.: Cloud interoperability and standardization. SETLabs Brief 7(7), 19–26 (2009)
24. Aceto, G., Botta, A., de Donato, W., Pescapè, A.: Cloud Monitoring: definitions, issues and future directions, In Cloud Networking, IEEE 1st International Conference (2012)
25. Tapp, http://www.tapp.in/ (2013)
26. Rodero-Merino, L., Vaquero, L.M., Gil, V., Galán, F., Fontán, J., Montero, R.S., Llorente, I. M.: From infrastructure delivery to service management in clouds. Future Gener. Comput. Syst. **26**(8), 1226–1240 (2010)
27. Aminullah, S., Molina-Jimenez, C.: Cost Estimation of Service Delivery in Cloud Computing (Dissertation, Arjuna Technologies Ltd.) (2012)
28. Plummer, D., Gartner Inc.: The Business Landscape of Cloud Computing, Financial Times (2012)
29. Buyya, R., Yeo, C.S., Venugopal, S.: Market-oriented cloud computing: vision, hype, and reality for delivering it services as computing utilities. In: High Performance Computing and Communications, HPCC'08. 10th IEEE International Conference (2008)

30. Dawson, C.J., et al.: Optimizing cloud service delivery within a cloud computing environment, US Patent Application 12/630,275 (2009)
31. Buyya, R., Ranjan, R., Calheiros, R.N.: Modeling and simulation of scalable Cloud computing environments and the CloudSim toolkit: challenges and opportunities. In: High Performance Computing & Simulation, HPCS'09. International Conference on. IEEE (2009)
32. Cloud Security Alliance: The Security, Trust & Assurance Registry (STAR). https://cloudsecurityalliance.org/star/ (2011)
33. Plummer, D., Gartner Inc: Cloud Service Brokerage: A Must-Have for Most Organizations, Forbes (2012)
34. Venkatraman, S., Wadhwa, B.: Cloud Computing A Research Roadmap in Coalescence with Software Engineering. SEIJ **2**(2), 7–18 (2012)

Hybrid Intelligent System for Lung Cancer Type Identification

Yenatfanta Shifferaw and Kumudha Raimond

Abstract Lung cancer is one of the deadly and most common diseases in the world. Most of the current research works are based only on classifying the nodules (tissue mass in the lungs) as cancerous or noncancerous (NC). In this work, a hybrid intelligent lung cancer identification system is proposed to identify the two general types of lung cancers such as small cell lung cancer (SCLC) and nonsmall cell lung cancer (NSCLC) using computed tomography (CT) images. The proposed system followed two approaches to extract the features from CT images and to identify the type using self-organizing map (SOM) and fuzzy logic concepts. The system is analyzed using 86 images. The first and second approach has resulted in 50.11 % and 97.22 % of accuracy, respectively.

Keywords Hybrid intelligent system • Lung cancer • Cancer identification

1 Introduction

Lung cancer is a disease [1] which consists of uncontrolled cell growth in tissues of the lungs. This growth may lead to metastasis, which is the invasion of adjacent tissue and infiltration beyond the lungs. The main types of lung cancer are SCLC and NSCLC. SCLC refers to malignant form of cancer and is characterized by small round or oval cells, whereas NSCLC applies to the various types of bronchogenic carcinomas. It is important to distinguish the type of cell as the treatment varies;

Y. Shifferaw (✉)
Department of Electrical and Computer Engineering, Addis Ababa Institute of Technology, Addis Ababa University, Addis Ababa, Ethiopia
e-mail: yenatshiff@yahoo.com

K. Raimond
Department of Computer Science and Engineering, Karunya University, Coimbatore, TN, India
e-mail: kraimond@karunya.edu

E.B. Rajsingh et al. (eds.), *Informatics and Communication Technologies for Societal Development*, DOI 10.1007/978-81-322-1916-3_5,

NSCLC is treated with surgery, while SCLC responds better to chemotherapy and radiation [2].

Lung cancer appears in the form of nodules or lesions in the lungs that are either solitary or attached to the lung wall, major vessels, or other structures. CT is the most sensitive imaging technique for detecting lung cancer at an early stage. However, the volumetric data acquired by CT scanners produces so many images that radiologists need an automated or semi-automated aid in interpreting the images more rapidly. Besides, the treatment requires quantitative estimate of lesion load which demands accurate identification of nodules. This is possible only with automated or semi-automated aid.

Generally, computer-aided analysis of medical images obtained from different imaging systems [3] involves four basic phases: image preprocessing, image segmentation, data reduction/feature extraction, and identification. Research is being carried out in each of the phases to enhance the outputs.

In [4], sputum color images are analyzed using Hopfield neural network (HNN) for detecting lung cancer. One thousand samples of sputum images are analyzed, and the proposed system and the pathologists have agreed on 92 % of the cases and disagreed on 8 % of the cases.

In [5], an automatic pathological diagnosis procedure named neural ensemble-based detection is proposed, which uses neural network ensemble to identify lung cancer cells in the specimens of needle biopsies obtained from the patients. It is able to achieve an overall accuracy of 88.4 %.

In [6], a computer-aided diagnosis system is developed to find the early lung cancer nodules and to classify them as benign or malignant. The work in [7] has proposed an automated pulmonary segmentation algorithm followed by nodule classification using rough set classifier. In [8], recent research works are reviewed where most of the works are based on nodule classification.

Most of the research works under "Nodule detection and classification" are limited to classify the lung as cancerous or NC, and further classification is made with sputum cytology, but the efficiency of this modality depends on the number of sputum samples, specimen adequacy, and highly trained personnel and it is also time consuming. The other method which is based on biopsy data involves the risk of infection and lung collapse. Besides, in both the approaches, CT image has to be analyzed first to find out whether there are suspicious regions or not.

After closely reviewing the reported works related to lung cancer detection, the proposed system is designed using hybrid intelligent systems to identify the two general types of lung cancers such as SCLC and NSCLC using CT images. The rest of this chapter is organized as follows. Section 2 describes the design of the proposed approach. Result analysis is covered in Sect. 3. Conclusions are summarized in Sect. 4.

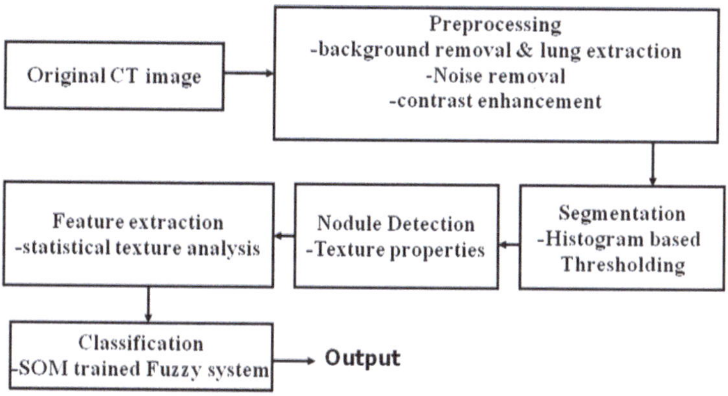

Fig. 1 Block diagram of the proposed approach

2 Design and Implementation

In this work, two approaches are followed to identify lung cancer; the first approach is the general way that most of the literatures follow to identify nodules. It includes preprocessing, nodule detection, feature extraction, feature selection, and classification phases. The features of each nodule are given as input for the classifier. However, in the proposed approach, the features are extracted from the lung image rather than from each nodule. All the other modules are same as the first approach. The block diagram representation of the first approach designed is shown in Fig. 1. The details of the system are explained below.

2.1 *Lung Image Database*

The NC images were taken locally from Inter-medical Diagnostic center whereas SCLC and NSCLC images were collected from the National Cancer Institute (NCI) called Lung Image Database Consortium (LIDC). The number of images collected for NSCLC is 28 from 6 patients, 18 for SCLC from 2 patients, and 40 for NC from 16 patients. Totally, 86 images are used in the system for training and testing. Figure 2 depicts the initial, middle, and last slices of NC, SCLC, and NSCLC images, respectively.

2.2 *Preprocessing*

Preprocessing is an important step (Fig. 3) to improve the results in the subsequent stages. Stages of preprocessing are shown in Fig. 3. The resultant images are shown in Fig. 4.

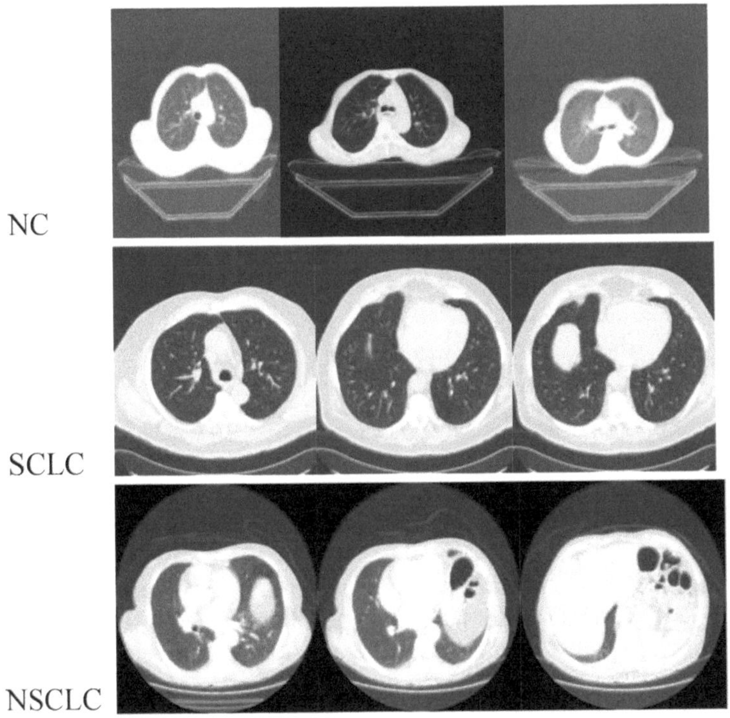

Fig. 2 Sample original images

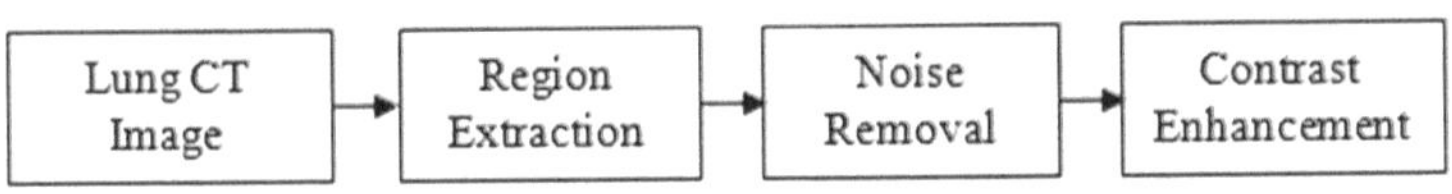

Fig. 3 Preprocessing stage

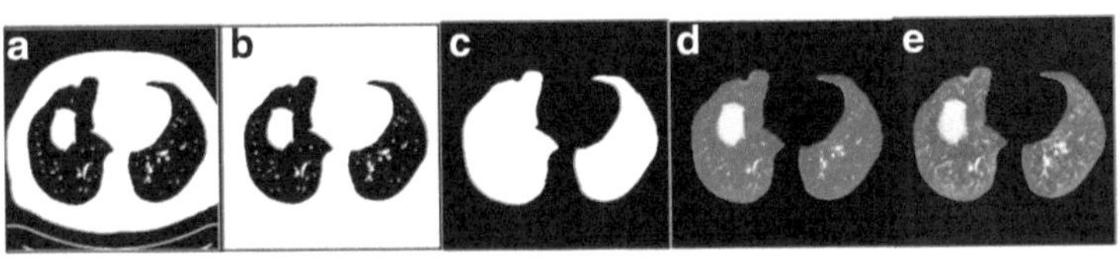

Fig. 4 Results of preprocessing stage

Region extraction is the first step under preprocessing, and it is used to retrieve the lung area by eliminating other parts of the chest image and CT apparatus image. The available approaches to extract the region of interest are thresholding, region growing, connected components labeling, morphological operations, rolling ball algorithm, flood-fill operations, and edge detection with their own advantages and disadvantages. And combining few of them will compensate their limitations. In this work, the following approaches are used to extract the lung region.

- *Thresholding*: The image is first converted to gray scale and then to binary image by replacing all pixels with luminance greater than the threshold with value 1 (white) and all other pixels with the value 0 (black). The threshold value is computed using Otsu's method and the result is shown in Fig. 4a.
- *Flood-Fill Operation*: In this work, 8-neighborhood connectivity is used to fill any hole with intensity values similar to the surrounding pixels [9]. This method is useful to remove the background of the image by subtracting the negative of the filled image from the binary image. The corresponding output is shown in Fig. 4b.
- *Edge Detection*: This step is to identify and locate abrupt changes in pixel intensity which characterizes the boundaries of objects in an image. Sobel, Prewitt, and Canny are some of the edge detection operators. Canny edge detector algorithm [10] is used in this work as it can detect edges effectively.
- *Connected Components Labeling*: This task identifies the objects and labels them using 8-connected neighborhoods and subsequently removes smaller objects and retains larger objects. Subsequently, filling and masking operations are applied and the resultant images are shown in Fig. 4c, d, respectively. Masking operation is used to get the extracted lung region by setting all the non-filled pixels of image (Fig. 4c) to zero.
- *Contrast Enhancement*: Median filter is used to remove the noises from the extracted lung region. The image is further enhanced using adaptive histogram equalization method [11] which is a spatial domain method and provides better result in enhancing low-illumination images, and the resultant image is shown in Fig. 4e.

2.3 Nodule Detection

Nodule detection is the second stage of the first approach and it involves two steps: segmentation and false nodule reduction.

- *Segmentation*: Segmentation of nontrivial images is one of the difficult tasks in image processing. The eventual success or failure of computerized analysis procedure depends on the accuracy of segmentation. In case of medical images, segmentation is useful to locate tumor, lesion, and other abnormalities. Segmentation can be carried out based on threshold, region, deformable models, fuzzy logic, and neural network [12]. In this work, bimodal histogram thresholding technique is used which can characterize the image histogram using two dominant modes [13]. The main objective of segmentation in this work is to identify the nodule candidates.

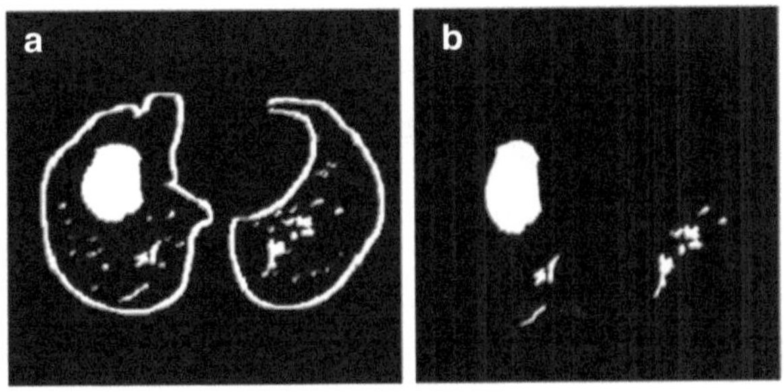

Fig. 5 Candidate nodules

- *False Nodule Reduction*: Some blood vessels and other nodule-like structures might be seen in the image. It is necessary to remove them from the image so as to identify the true nodules. The results are shown in Fig. 5a, b.
- *Feature Extraction for Nodule Detection*: This step involves extracting features from each nodule so as to distinguish true from false nodules. The diameter and the circularity of each nodule are calculated based on the formula followed. For this work, diameter and circularity of size ≥2 mm and ≥0.2 mm, respectively, are chosen to characterize a nodule.

2.4 Feature Extraction

Features are extracted from candidate nodules in this phase. Different features (Fig. 6) representing each nodule are calculated and averaged to get the features of the whole image. Figure 6 shows the types of features extracted from each preprocessed image.

Mean, standard deviation, smoothness, and skewness are calculated as first-order statistical features [14]; contrast, homogeneity, correlation, entropy, uniformity, and energy are obtained as second-order statistic features [15] for each of the four directions and averaged to get the final result. The geometric features obtained are area, perimeter, and diameter. Higher-order statistical features are short-run emphasis, long-run emphasis, gray-level nonuniformity, run length nonuniformity, run percentage, low gray-level run emphasis, high gray-level run emphasis, short-run low gray-level emphasis, short-run high gray-level emphasis, long-run low gray-level emphasis, and long-run high gray-level emphasis [16]. Totally, 24 features are given to the feature selection phase after normalization.

2.5 Feature Selection

Multinomial logistic regression method is used to determine the most relevant features. Through this method, 9 features got selected out of the 24 features. Dataset

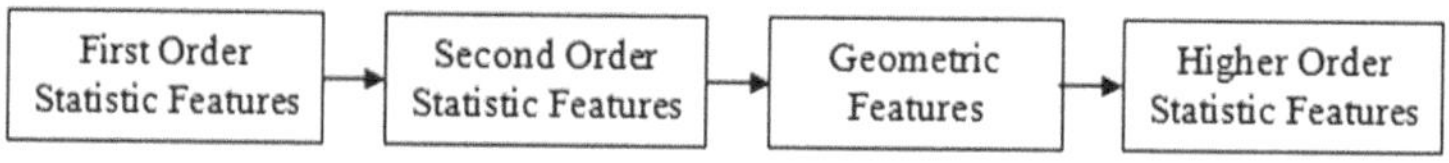

Fig. 6 Components of feature extraction phase

of high variance are used for training and the rest for testing purpose. The calculation has yielded 50 images for training, out of which 5 are SCLC, 22 NSCLC, and 23 NC, and 36 for testing, in which 13 are SCLC, 6 NSCLC, and 17 NC images.

2.6 *Classification*

The classification module consists of two steps: clustering and identification. SOM is used to cluster the dataset. The number of input and output and initial number of neighborhood are kept constant, whereas the influence of learning rate and the total number of iterations on accuracy are studied and the results are shown in Tables 2 and 3, respectively. The final weight values from SOM are given to fuzzy system. The identification process is based on fuzzy logic [17]. The stopping criterion was set to root mean square error of value 10^{-6}. It has 8 input nodes which is the output of the SOM and the system as a whole has two output nodes. The learning rate for the fuzzy logic is set to 1 and the threshold value for the output function is set to 0.5. The output of the fuzzy logic is mapped into three values which are 00 for NC, 01 for SCLC, and 10 for NSCLC.

2.7 *Proposed Approach*

The preprocessing phase is common to both the approaches. However, in the next phase, 24 features are extracted from the whole image rather than from the nodules. Of these, nine features which showed high predictor value for the desired output are selected for the classification purpose.

3 Performance Analysis

The performance is measured using accuracy, sensitivity, and specificity. Table 1 shows the accuracy of preprocessing and nodule detection stages. Tables 2 and 3 show the performance metrics of hybrid intelligent system of the first and second approach, respectively. It is clear from the results that the proposed approach is better than the first approach.

Table 1 Accuracy of preprocessing and nodule detection stage

	Accuracy (%)	
Cancer type	Preprocessing stage	Nodule detection
NC	100	95
NSCLC	89.28	78.57
SCLC	100	88.88

Table 2 Accuracy of classification stage of the first approach

	Parameters					
	Learning rate: 0.01	Learning rate: 0.1	Learning rate: 0.01	Learning rate: 0.01	Learning rate: 0.01	Learning rate: 0.01
Performance metrics	Epoch: 50	Epoch: 50	Epoch: 100	Epoch: 150	Epoch: 150	Epoch: 150
Accuracy	36.11	27.77	22.22	*50.11*	19.44	22.22
Sensitivity	25.34	21.84	20.96	*54.89*	17.89	14.67
Specificity	30.26	23.65	32.13	*43.24*	23.24	20.78

Table 3 Accuracy of classification stage of the second approach

	Parameters					
	Learning rate: 0.01	Learning rate: 0.1	Learning rate: 0.01	Learning rate: 0.01	Learning rate: 0.01	Learning rate: 0.01
Performance metrics	Epoch: 50	Epoch: 50	Epoch: 100	Epoch: 150	Epoch: 100	Epoch: 100
Accuracy	83.33	75	*97.22*	13.43	11.11	50.2
Sensitivity	100	93.33	*92.86*	100	9.8	34.67
Specificity	72.72	61.9	*100*	61.11	34.26	28.59

4 Conclusion

The system has been experimentally analyzed using 86 images, in which 50 are used for training and 36 for testing. The first approach presents an identification capacity with accuracy of 50.11 %, and the second approach has given accuracy of 97.22 %.

References

1. World Health Organization, Cancer. http://www.who.int/mediacentre/factsheets/fs297/en/ (18 Feb 2014)
2. Vaporciyan, A.A., Nesbitt, J.C., Lee, J.S.: Cancer medicine. B C Decker, Hamilton (2000)
3. Mughal, M.N., Ikram, W.: Early lung cancer detection by classifying chest CT images: a survey. In: 8th International Multitopic Conference, Proceedings of INMIC, pp. 67–72. (2004)
4. Taher, F., Sammouda, R.: Morphology analysis of sputum color images for early lung cancer diagnosis. In: 10th International Conference on Information Sciences Signal Processing and their Applications, pp. 296–299. (2010)

5. Zhou, Z.-H., Jiang, Y., Yang, Y.-B., Chen, S.-F.: Lung cancer cell identification based on artificial neural network ensembles. Artif. Intell. Med. **24**(1), 25–36 (2002)
6. Sharma, D., Jindal, G.: Identifying lung cancer using image processing techniques. In: International Conference on Computational Techniques and Artificial Intelligence (ICCTAI'2011), pp. 115–120. (2011)
7. Wong, T.-Y., Cheng, C.-H.: Automated segmentation and hybrid classifier for identifying medical image. J. Signal Process. Image Process. Pattern Recognit. **6**(1), 191–202 (2013)
8. Darne, K.S., Panicker, S.S.: Use of fuzzy C-mean and fuzzy min-max neural network in lung cancer detection. Int. J. Soft Comput. Eng. (IJSCE) **3**(3), 265–269 (2013)
9. Flood-fill operation. http://www.mathworks.in/searchresults/?c%5B%5D=entiresite&q=flood+fill (18th Feb 2014)
10. Juneja, M., Sandhu, P.S.: Performance evaluation of edge detection techniques for images in spatial domain. Int. J. Comput. Theor. Eng. **1**(5), 614–621 (2009)
11. Adaptive histogram equalization. http://en.wikipedia.org/wiki/Histogram_equalization (18th Feb 2014)
12. Sharma, N., Aggarwal, L.M.: Automated medical image segmentation techniques. J. Med. Phys. **35**(1), 3–14 (2010)
13. Prasantha, H.S., Shashidhara, H.L., Murthy, K.N.B., Lata, G.M.: Medical image segmentation. Int. J. Comput. Sci. Eng. **2**(4), 1209–1218 (2010)
14. Srinivasan, G.N., Shobha, G.: Statistical texture analysis. Proc. World Acad. Sci. Eng. Technol. **36**, 1264–1269 (2008)
15. Mokji, M.M., Bakar, S.A.R.A: Gray level co-occurrence matrix computation based on Haar wavelet. In: Computer Graphics, Imaging and Visualization, pp. 273–279. (2007)
16. Kelvin Law: Detection of disease in CT Scans of the lung using texture analysis and machine learning techniques. Project Report submitted in support of the degree of Master of Engineering, University of Bristol (2004)
17. Mu-Chun Su, Chee-Yuen Tew: A self-organizing feature-map-based fuzzy system. In: International Joint Conference on Neural Networks, IJCNN 2000, Proceedings of the IEEE-INNS-ENNS, vol. 5, pp. 20–25. (2000)

Auditing Mechanism to Verify the Integrity of Remote Data in Cloud

V. Nirmala, R. Shanmugalakshmi, and M. Kiruthika

Abstract Cloud computing is the one which offers infrastructure to the client on a pay-as-you-use model by leveraging on the Internet technology. As this model depicts the electricity billing, it is often termed as utility computing. As the use of cloud increases day by day, the concerns about cloud also increase. There are a number of data breaches that are happening day by day that affect the cloud growth. The major concern in adapting cloud is its security. In order to solve the security issues, there are a number of solutions proposed. This chapter focuses on the integrity assurance strategy for remote data. The technique proposed combines the encrypting mechanism along with the data integrity check mechanism. The analysis and simulation results show that our proposed auditing protocols are secure and it reduces the computation load and communication cost of the auditor.

Keywords Storage auditing • Integrity • Bilinear pairing • Cloud storage • Cloud computing

1 Introduction

Cloud computing is a term for anything that involves delivering services over the Internet. Cloud storage is an essential service of cloud computing [8–10], which allows data owners to move data from their local computing systems to the cloud. An increasing number of individuals and organizations started placing their data in the cloud. There are a number of challenges to be faced while placing the data in the cloud. When the data is static, a simple hash value would be enough to verify data safety. If the data is dynamic, then simple encryption and hash value verification will not help. There are solutions proposed recently to work on the data while the

V. Nirmala (✉) • R. Shanmugalakshmi • M. Kiruthika
Department of CSE/IT, Government College of Technology, Coimbatore, TN, India
e-mail: nimmicsegct@gmail.com; cseit.gct@gmail.com; kiruthi.skm@gmail.com

E.B. Rajsingh et al. (eds.), *Informatics and Communication Technologies for Societal Development*, DOI 10.1007/978-81-322-1916-3_6, © Springer India 2015

data is being encrypted. The data owner would worry that the data could be lost in the cloud. There are chances that the data placed in the cloud can be lost due to hardware malfunction or few of the service providers would intentionally delete the data to satisfy an immense number of customers.

The third-party auditors are the skilled persons in a particular verification field. The owner can employ those third-party auditors to verify their data by providing the necessary information for auditing. In order to bring the users to the cloud and to make them feel secure of their data, the service providers should allow the auditors to verify the owner data.

2 Existing System

There are a number of integrity verification protocols which have been proposed in recent times. The proposed methods use different techniques to enhance data privacy for remote data. The PDP [1] model uses the hash method to verify data privacy. When data gets updated, all hash values have to be recomputed and are used only for static data. The public auditability [2] scheme uses the Merkle hash tree data structure for data privacy but may reveal the data to the auditor. The POR [3] scheme suggests the BLS method for data privacy but does not support batch auditing for multiple clouds. Table 1 gives the comparisons among some existing remote integrity checking schemes in terms of the type of data, the privacy protection, the support of dynamic operations, and the batch auditing for multiple owners and multiple clouds.

3 Proposed System

3.1 System Model

The proposed model consists of three entities: data owner, auditor, and cloud server. Initially the data owner places the data in the cloud by trusting the service provider. The service provider stores data on behalf of the data owner and safeguards it. To verify data integrity, the owner employs the auditor with necessary information for performing the audit operation over the data. The auditor takes charge of verifying the data and sends the verified proof to the owner. Figure 1 shows the system model for the proposed scheme which incorporates three entities. The owner receives the acknowledgment from the server for placing the data and verification proof from the auditor after the auditor has performed the verification/ periodic auditing.

Table 1 Comparison of different integrity verification techniques

Method	Technique used	Type of data	Privacy	Dynamic operations	Batch operations
PDP [1]	Homomorphic verifiable tag	Static data	Yes	Not allowed	Not supported
POR [3]	Sentinels	Static data	Yes	Not allowed	Not supported
Public auditability [2]	Merkle hash trees and bilinear aggregation function	Dynamic data	Yes	Allowed	Supported
POR scheme with full-proof mechanism [5]	Homomorphic authentication technique with BLS signature	Dynamic data	Yes	Allowed	Not supported
Cooperative PDP model [6]	Homomorphic verifiable response and hash index hierarchy	Dynamic data	No	Not allowed	Not supported
Dynamic PDP model [7]	Rank-based authentication	Dynamic data	No	Allowed but partially	Not supported
Our scheme	Bilinear pairing	Dynamic data	Yes	Allowed	Supported

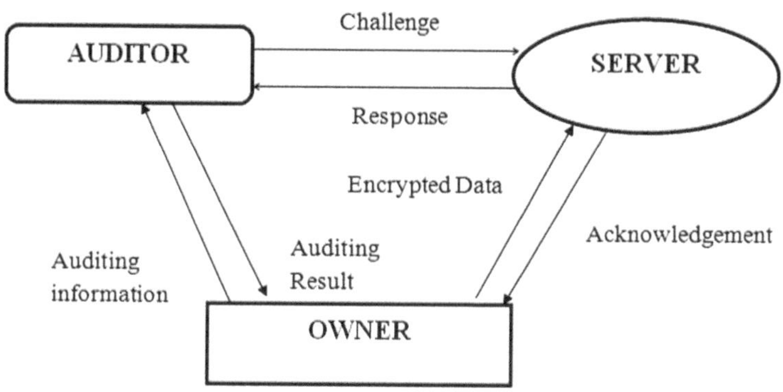

Fig. 1 System model

4 Integrity Assurance Protocol

The integrity assurance protocol is proposed based on bilinear assumption. The proposed system is used for outsourced data privacy. The algorithm, flow, and the auditing mechanism are proposed below.

4.1 Problem Definition

The work provides a solution for the data privacy problem in the cloud environment by introducing a third-party auditor to audit and preserve the integrity of the data.

4.2 System Setup

A file F is divided into a number of data blocks. Each data block is allocated with a tag. The tag is used to generate proof for the data, without having any knowledge over the data [11, 12].

The auditor initially poses the challenge over the server for initial verification for the entire data [14]. The result is maintained by the auditor and the copy is sent to the owner. Whenever there is a change to be done on the data, the owner updates the data in the cloud. At the same time, the owner sends the updated details to the auditor. The auditor performs periodic auditing based on the information given by the owner.

Bilinear Pairing: Let G1, G2, and GT be three multiplicative groups with the same prime order p [4].

A bilinear map is a map e: G1 $\times$ G2 $\rightarrow$ GT. Let g1 and g2 be the generators of G1 and G2, respectively. A bilinear map has the following properties:

1. Bilinearity: e (ua, vb) = e (u, v) ab for all u $\in$ G1, v $\in$ G2, and a, b $\in$ Zp.
2. Non-degeneracy: There exist u $\in$ G1, v $\in$ G2 such that e (u, v) = I, where I is the identity element of GT.
3. Computability: e can be computed in an efficient way.

4.3 Integrity Assurance Flow

The integrity assurance flow is divided into two stages. During the preprocessing stage, the owner generates keys (secret hash key (shk), secret tag key (stk), public tag key (ptk)) and a set of tags T for each block based on the security parameter. The server stores the data and their corresponding tags sent by the owner. The owner sends the key (shk, ptk) and abstract information about the data to the auditor. Based on the information received from the owner, the auditor will conduct the audit on the stored data (Figs. 2 and 3).

In the verification stage, the auditor challenges $\rho = (\{i, Vi\}, r)$ the server to verify the data using challenge stamp CS. The server creates the response R = (*DR*, *TR*) (response based on data *DR*, response based on tag *TR*) and sends it back to the auditor. The auditor verifies the data based on the challenge C, the response R, the secret hash key, the public tag key, and the abstract information of the data component. It first computes the hash values h (shk, block identifier Yi) of all the challenged

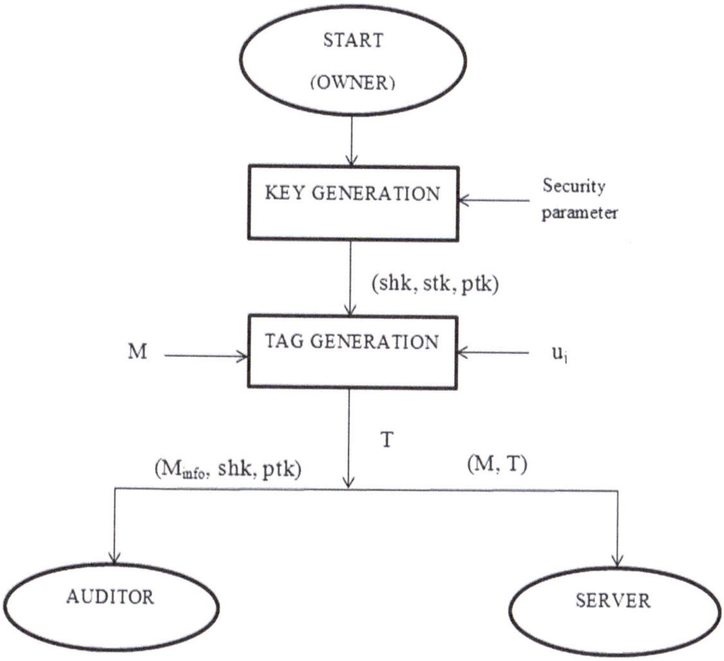

Fig. 2 Preprocessing stage

data blocks and computes the challenge hash chall and then verifies the response from the server by [15]

$$\prod DR.e(chall, ptk) = e\left(TR, g1^{rand}\right) \tag{1}$$

The verification is true when the algorithm outputs 1. Otherwise, if the output is 0, the data is unsafe [16].

$$\mathbf{Verify}\ (\mathbf{CS}, \mathbf{R}, \mathbf{shk}, \mathbf{ptk}, \mathbf{Minfo}) \rightarrow \mathbf{0/1}$$

5 Protocol Construction

During the system initialization, the owner generates the keys and the tags for the data. After storing the data on the server, the owner asks the auditor to conduct the verification auditing to make sure that their data is correctly stored on the server. Once confirmed, the owner can choose to delete the local copy of the data. Then, the auditor conducts the periodic auditing to check the data integrity [13].

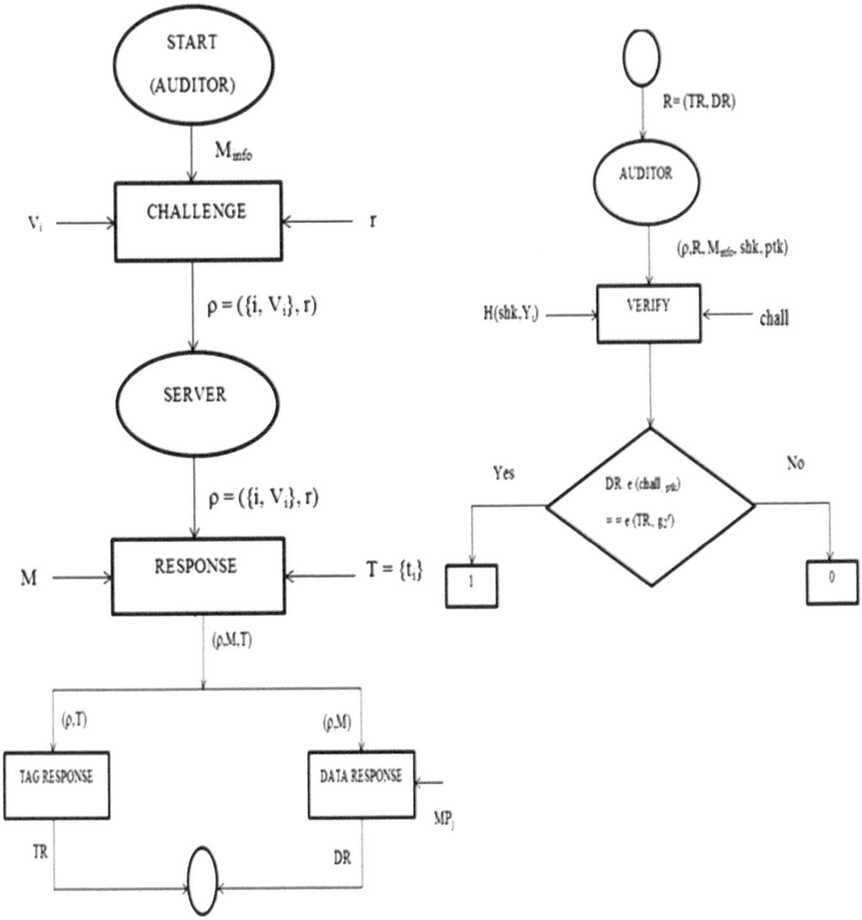

Fig. 3 Verification stage

5.1 Initialization

In this phase, the owner generates the key and tag before placing the data in the cloud. The pseudocode is given below.

1. The data will be encrypted.
2. The encrypted data will be divided into n blocks.
3. Generate key (secret hash key, secret tag key, public tag key) based on the bilinear pairing.
4. Generate the set of tags.
5. Attach the tag for each block.
6. Send the data and their corresponding tags in the cloud.
7. Send the data abstract along with the secret hash key and public tag key to the auditor.

5.2 Verification Auditing

The verification auditing protocol involves a challenge-response mechanism to audit the entire contents stored in the server. During this phase, the owner requires the auditor to check whether the owner's data is correctly stored on the server, and the steps involved are given below.

Auditor:

1. Save the data abstract along with the secret hash key and public tag key sent by the owner.
2. Challenge is generated based on the abstract information about the data for all data blocks.
3. Send the challenge to the server.
4. Based on the response received from the server, the auditor will start verification.
5. **If** Verify = 0, **then** the data is safe **else** the data is modified.
6. Trigger the message to the owner.

Server:

1. Store the data in the cloud sent by the owner.
2. The server will prepare the response for the challenge it received from the auditor.
3. Response is generated based on the data, tag, and challenge.
4. Send the response to the auditor.

5.3 Periodic Auditing

The periodic auditing will be carried out on a set of trial data blocks. The auditing takes on based on the owner's request to the auditor and on the service-level agreements signed between the data owner and the server.

Auditor:

1. Challenge is generated based on the abstract information about the data only for the trial data block.
2. Send the challenge to the server.
3. Starts periodic auditing based on challenge, secret hash key, public tag key and response it received from the server.
4. **If** Verify = 0, **then** the trial data is safe **else** the trial data is modified.
5. Trigger the message to the owner.

Server:

1. Based on the challenge it receives for the trial data, the server will find the response.
2. Response is generated for trial data based on the data, tag, and challenge.
3. Send the response to the auditor.

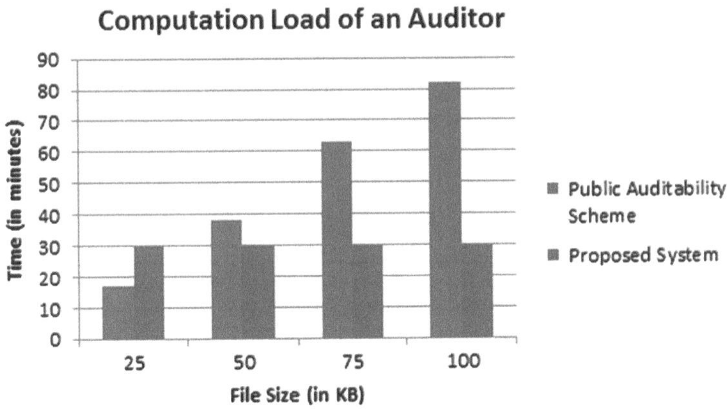

Fig. 4 Comparison of computation load

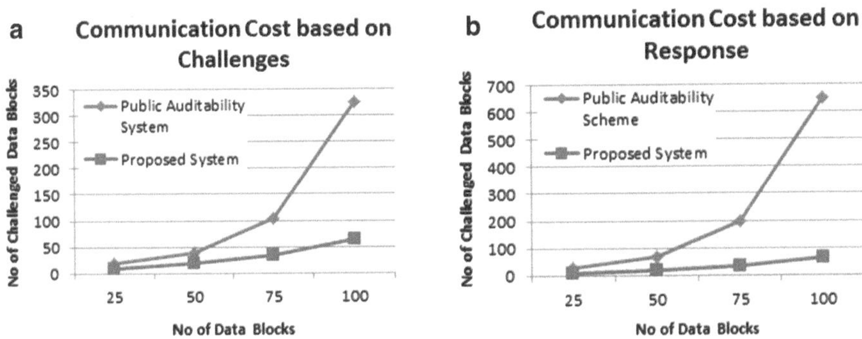

Fig. 5 Comparison of communication cost. (**a**) Based on the number of challenges. (b) Based on the number of responses

6 Result Analysis

The simulation is done on a Windows system with an i5 CPU at 2.00 GHz and 4GB RAM. The implementation uses PBC Library version 0.5.14. We compare the computation load of the auditor on the basis of the file size in KB and time in minutes. In our proposed model, the auditing time remains constant for varying file sizes, whereas the auditing time varies with the file size for the public auditability system (Fig. 4).

We compare the communication cost between the auditor and the server based on the number of challenges and responses. In our proposed system, the data is divided into n blocks, but in the existing system, each block is again divided into s sectors. So the communication cost is gradually reduced compared to the public auditability scheme. The data which are to be placed in the cloud are initially encrypted using the key which is generated by using the bilinear generators. As the proposed scheme is based on the assumption of the bilinear mapping scheme, the scheme is assumed to provide better data security than other existing techniques (Fig. 5).

7 Conclusion

Our auditing protocol ensures data privacy by using the encryption technique along with the bilinear property of the bilinear pairing. In our method, the server computes the proof as an intermediate value of the verification, such that the auditor can directly use this intermediate value to verify the correctness of the data. Therefore, the computation load of an auditor was greatly reduced which in turn improves the auditing performance. The auditing protocol can be extended to support the data dynamic operation and batch auditing for multiple owners as well as multiple clouds. The proposed scheme provides better security along with reduced computation and communication cost.

References

1. Ateniese, G., Burns, R.C., Curtmola, R., Herring, J., Kissner, L., Peterson, Z.N.J., Song, D.X.: Provable data possession at untrusted stores. In: ACM Conference on Computer and Communications Security, pp. 598–609, ACM, Alexandria, VA, USA (2007)
2. Shah, M.A., Swaminathan, R., Baker, M.: Privacy-Preserving Audit and Extraction of Digital Contents, 2008, Report 2008/186, Cryptology ePrint Archive
3. Juels, A., Kaliski, B.S. Jr.: Pors: proofs of retrievability for large files. In: Proceedings of the 14th ACM Conference on Computer and Communications Security (CCS'07), pp. 584–597, Alexandria, VA, USA (2007)
4. http://courses.csail.mit.edu/6.897/spring04/L25.pdf
5. Chang, E.C., Xu, J.: Remote integrity check with dishonest storage server. In: Proceedings of the 13th European Symposium Research in Computer Security (ESORICS'08), pp. 223–237, Malaga, Spain (2008)
6. Zhu, Y., Hu, H., Ahn, G., Yu, M.: Cooperative provable data possession for integrity verification in multi-cloud storage. In: IEEE Transactions on Parallel and Distributed Systems, pp. 2231–2244. (2012)
7. Erway, C., Kupcu, A., Papamanthou, C., Tamassia, R.: Dynamic provable data possession. In: Proceedings of the 16th ACM Conference on Computer and Communications Security (CCS'09), (2009)
8. Mell, P., Grance, T.: The NIST definition of cloud computing. National Institute of Standards and Technology, Tech. Rep., (2009)
9. Armbrust, M., Fox, A., Griffith, R., Joseph, A.D., Katz, R.H., Konwinski, A., Lee, G., Patterson, D.A., Rabkin, A., Stoica, I., Zaharia, M.: A view of cloud computing. Commun. ACM **53**(4), 50–58 (2010)
10. Velte, T., Velte, A., Elsenpeter, R.: Cloud Computing: A Practical Approach, 1st edn. McGraw-Hill, Inc., New York (2010)

11. Li, J., Krohn, M.N., Mazières, D., Shasha, D.: Secure untrusted data repository (sundr). In: Proceedings of the 6th Conference on Symposium on Operating Systems Design & Implementation, pp. 121–136, Berkeley, CA, USA (2004)
12. Lillibridge, M., Elnikety, S., Birrell, A., Burrows, M., Isard, M.: A cooperative internet backup scheme. In: USENIX Annual Technical Conference, General Track. USENIX, pp. 29–41. (2003)
13. Shacham, H., Waters, B.: Compact proofs of retrievability. In: Pieprzyk, J. (eds.) ASIACRYPT.Lecture Notes in Computer Science, vol. 5350, pp. 90–107. Springer, Melvourne, Australia (2008)
14. Erway, C.C., K¨upc,¨u, A., Papamanthou, C., Tamassia, R.: Dynamic provable data possession. In: Al-Shaer, E., Jha, S., Keromytis, A.D. (eds.) ACM Conference on Computer and Communications Security, pp. 213–222, ACM, Chicago, IL, USA (2009)
15. Kan Yang, Xiaohua Jia: An efficient and secure dynamic auditing protocol for data storage in Cloud computing. IEEE Trans. Parallel Distrib. Syst. **24**(9), 1–11 (2012)
16. Zhu, Y., Wang, H., Hu, Z., Ahn, G.-J., Hu, H., Yau, S.S.: Dynamic audit services for outsourced storages in clouds. IEEE Trans. Serv. Comput. **6**(2), 227–238 (2013)

An Efficient and Secure Location Update Protocol Using Fractal-Based Public Keys in Mobile IP Networks

Senthilkumar Mathi and M.L. Valarmathi

Abstract The process of informing the current location of a mobile device to its home network after moving to another network is called a location update or registration in mobile IP networks. This process must be secured as the registration messages can be intercepted and falsified by an intruder. The security provided must also have minimal registration latency, as both are directly proportional. Therefore, it is necessary to design a registration with balanced effort between security and registration latency. Hence, this chapter deals with a new proposal to balance both using fractal-based scheme, which is a substitute to conventional number theory-based schemes. The security properties of the proposed protocol are evaluated and validated using a model checker AVISPA. The performance analysis of the proposed protocol and comparison of numerical results with existing protocols shows that our protocol surpasses the existing protocols in terms of balanced security and registration latency.

Keywords Authentication • Confidentiality • Fractal • IP mobility • Registration latency • User anonymity

S. Mathi (✉)
Department of Computer Science and Engineering, Amrita School of Engineering, Amrita Vishwa Vidyapeetham University, Coimbatore, TN, India
e-mail: m_senthil@cb.amrita.edu

M.L. Valarmathi (✉)
Department of Computer Science and Engineering, Government College of Technology, Coimbatore, TN, India
e-mail: ml_valarmathi@rediffmail.com

E.B. Rajsingh et al. (eds.), *Informatics and Communication Technologies for Societal Development*, DOI 10.1007/978-81-322-1916-3_7, © Springer India 2015

1 Prologue

The increasing variety of wireless devices offering Internet protocol (IP) connectivity is established to change the perceptions of the Internet as per the requirement of the current technology. A mobile host may be in use continuously through a wireless network interface as it is being carried from one place to another or it may simply be disconnected from one network and plugged into another network at a new location. On the other hand, present Internet routing protocols require the network address to change when the mobile host moves to the new location. This is tedious for laptops and other mobile hosts. The mobile IP (MIP) handles this problem using mobility agents [1]. In MIP registration model as shown in Fig. 1, a mobile node (MN) leaves from its home network to a foreign network. Afterwards, the MN gets a care-of-address (CoA) with a foreign agent (FA) in the foreign network through an advertisement message issued by the FA surrounded by the coverage area. Then, the MN sends a registration request 1 (R_Req1) message to FA. The FA then in turn sends the registration request 2 (R_Req2) message notifying a home agent (HA) to validate MN's request message. After the successful validation of the request message by HA, it updates the CoA of MN and replies FA with the registration reply 1 (R_Rep1) message. Then, the FA forwards the registration reply message 2 (R_Rep2) to MN. If any packets are destined to MN, then these are routed to FA by HA. Subsequently, the FA sends them to the MN.

Numerous studies [2] have been explored on the registration of MIP for security and/or efficiency issues, and consequently, different protocols have been suggested. However, this chapter proposes a new registration protocol based on fractal scheme to improve security and to reduce the computational cost. The security properties of the proposed protocol are validated using Automated Validation of Internet Security Protocol and Applications (AVISPA). Also, the protocol evaluation results have been experimented along with the comparative performance analysis.

2 Related Works

Initially, a base protocol [3] makes use of either timestamp or nonce. It uses secret keys and distributes them manually for the mutual authentication between the communicants. The advantage of this protocol is the lowest registration delay compared with other proposals and the prevention of replay attack through timestamp or nonce. But, it suffers from the drawback that it is not scalable due to the manual key distribution. To bring the scalability, the protocol [4] proposed by Jacobs provides the central authority that is responsible for distributing the key with the public key infrastructure (PKI) concept. It uses the certificate and digital signature for authentication. However, due to the MN's limited computing power, the Jacobs's proposal is not suitable for performing heavy PKI-based operations on

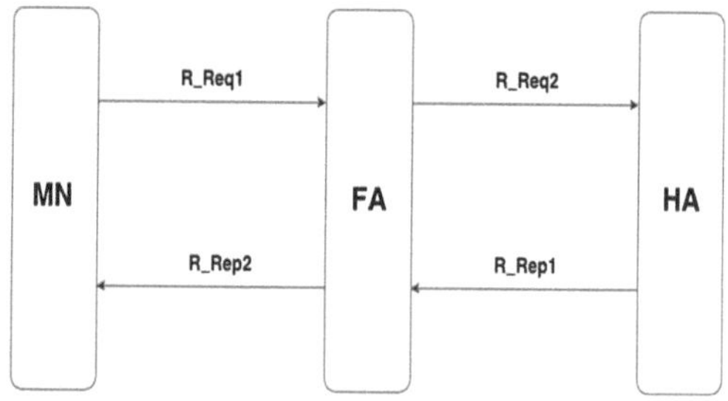

Fig. 1 Mobile IP registration environment

MN. To minimize the computing power requirement on MN, the minimal public key-based protocol [5] is proposed by Sufatrio and Lam that employs the minimal use of public keys and combines shared secret key and public key for authentication. Nevertheless, it does not provide the confidentiality and user anonymity between the MN and mobility agents. In order to bring the confidentiality between all correspondent pairs, Yang's protocol [6] is proposed with enciphering of messages between the correspondents by using temporary session key. However, the cryptographic operations of the Yang's protocol lead to high registration delay and there is no authentication between MN and FA. Subsequently, self-certified public key-based protocols [7] were proposed in terms of efficiency and security aspects with two different proposals, namely, time-invariant (TIV) and time-variant (TV) protocols. These protocols include the features such as parameter resynchronization, user anonymity, and prevention of replay attack. Even so, the registration time of two different variations is high due to the cryptographic operations. Subsequently, a protocol [8] is proposed, using certificateless PKI (CL-PKI) scheme. It attains the reduction in registration delay up to 30.47 ms. Further, the proposal given in [9] minimizes the registration delay with the minimal usage of certificateless public key signature. Next, the ID-based schemes [10, 11] were proposed based on an identity procedure that eliminates expensive pairing operations. These schemes reduce the registration latency. However, the extreme security measures in the ID-based schemes may possibly cause long registration time. Subsequently, to preserve the user anonymity in registration, the privacy-preserving registration protocol [12] is proposed for mobile network. Subsequently, other proposals [13, 14] are investigated for the location update with balanced effort between security and efficiency in terms of registration latency. Hence, the entities of the MIP should trust each other and the registration process must be secured against any sort of attacks. At the same time, the inclusion of excessive security includes long time for the registration. Therefore, we present a new registration proposal using fractal-based scheme to provide enhanced security and minimal latency between the agents without the third party.

Table 1 Notation used in the proposed protocol

Notation	Description
MN_{CoA} and MN_{tid}	MN's care-of-address temporary identity
HA_{id} and FA_{id}	Identity of HA and FA
MN_{Nonce}, HA_{Nonce}, FA_{Nonce}	Nonces of MN, HA, and FA, respectively
K_{MN-HA}, K_{FA-HA}	Secret key of MN and HA and FA and HA
Z_nFA, Z_kHA	Public key of FA and HA, respectively
$\{F_1, F_2 \ldots F_n\}K$	Encryption of fields $F_1, F_2 \ldots F_n$ using key K
$EText_n$	Encrypted text of nth message
H(P) and H^*	Hash value of P and computed hash value at receiver side
$F_1 \parallel F_2$	Concatenation of two fields F_1 and F_2
Request and Reply	Sequence of bits indicating a registration request and a reply
Requesttoforward	Sequence of bits indicating a request to forward message
Requesttoreply	Sequence of bits indicating agent to acknowledge the message
Announce	Sequence of bits indicating the presence of an agent
$A \rightarrow B$: $F_1, F_2 \ldots F_n$	An entity A sends $F_1, F_2 \ldots F_n$ to entity B

3 Proposed Location Update Protocol

The proposed protocol uses the notations shown in Table 1 to illustrate the message sequence.

The public keys of FA and HA are generated using Mandelbrot and Julia fractal functions [15]. The fractals are cross-sectional geometric shape that is produced from identical components that are small in size [16]. In the setup of our fractal-based scheme, FA and HA agree on global values *g*, *x* and use the private keys *(E, n)* and *(D, k)*, respectively. The values *E* and *D* are complex numbers belonging to Mandelbrot set. The values *n*, *k*, and *x* are random integers, where $x < k, n$. The public keys of the FA and HA are computed as follows and pre-shared through secure channel:

$$Z_nFA = g \times f(Z_{n-1}FA) \tag{1}$$

$$f(Z_{n-1}FA) = (Z_{n-1}FA) \times g \times E, \;\; Z_0 = g. \tag{2}$$

$$Z_kHA = g \times f(Z_{k-1}HA); \tag{3}$$

$$f(Z_{k-1}HA) = (Z_{k-1}HA) \times g \times D, \;\; Z_0 = g. \tag{4}$$

The proposed registration consists of the following steps:

1. FA $\rightarrow$ MN: P_1 where P_1 = Announce, MN_{CoA}, FA_{id}, FA_{Nonce}
2. MN $\rightarrow$ FA: P_2, H ($P_2 \parallel FA_{Nonce}$), $EText_1$ where P_2 = Requesttoforward, FA_{id}, MN_{tid}, HA_{id}. $EText_1$ = {Request, MN_{CoA}, HA_{id}, FA_{id}, HA_{Nonce}, MN_{Nonce}} K_{MN-HA}
3. FA (upon receipt of the step 2 message)

Calculate a new hash value H^* on the received fields of P_2 along with FA_{Nonce} and compare it with the received hash for authentication. Also, compute the secret key:

$$K_{FA-HA} = g^{n-x} \times (Z_k HA)_n E \tag{5}$$

4. FA → HA: P_3, H ($P_3 \| Z_n FA$), $EText_1$, $EText_2$ where P_3 = Requesttoreply, HA_{id}, FA_{id}, MN_{tid}. $EText_2 = \{HA_{id}, FA_{id}, FA_{Nonce}\}K_{FA-HA}$
5. HA (upon receipt of the step 4 message)

   Calculate a new H^* on the received fields of P_3 with public key $Z_n FA$ and compare it with the received hash value for authentication. Compute the secret key:

$$K_{HA-FA} = g^{k-x} \times (Z_n FA)_k D \tag{6}$$

   Decrypt $EText_2$ using K_{FA-HA} and validate FA_{Nonce} and decrypt $EText_1$ using K_{MN-HA} and validate the nonces of HA and MN. Store the new mobility binding MN_{CoA}. Generate new nonces HA'_{Nonce}, MN'_{Nonce}, and session key K'_{MN-HA} via a hash-based message authentication code (MAC) using secure hash algorithm, HMAC-SHA1,

$$K'_{MN-HA} = \text{HMAC-SHA-1}\left(K_{MN-HA}, HA'_{Nonce} \| MN'_{Nonce} \| HA_{id}\right) \tag{7}$$

6. HA → FA: P_4, H ($P_4 \| Z_n HA$), $EText_3$, $EText_4$ where P_4 = Requesttoforward, FA_{id}, HA_{id}, MN_{tid}, FA_{Nonce}. $EText_3 = \{Reply, HA_{id}, HA'_{Nonce}, MN'_{Nonce}, K'_{MN-HA}\}K_{FA\text{-}HA}$ and $EText_4 = \{HA_{id}, FA_{id}, HA_{Nonce}\}K_{HA-FA}$
7. FA (upon receipt of the message in step 6)

   Compute a new hash value H^* on the received fields of P_4 with the pre-shared public key $Z_n HA$ and compare it with the received hash value for authentication. Decrypt $EText_4$ using K_{HA-FA} and validate HA_{Nonce}.
8. FA → MN: P_5, H ($P_5 \| FA_{Nonce}$), $EText_3$ where $P_5 = FA_{id}$, MN_{tid} and FA_{Nonce}
9. MN (upon receipt of the message in step 8)

   Compute a new hash value H^* on the received fields of P_5 along with FA_{Nonce} and compare it with the received hash value for authentication. Decrypt the $EText_3$ using $K_{MN\text{-}HA}$ and recover the new nonces of HA, MN, and session key. Store HA'_{Nonce}, MN'_{Nonce}, and K'_{MN-HA} for the next registration.

4 Security Analysis

The delivery of data in the Internet can be easily interrupted and falsified. Thus, ensuring confidentiality between the source and destination is significant in MIP. In the proposed protocol, the secrecies of the necessary messages are encrypted using K_{MN-HA} and K_{FA-HA}. It is observed that the privacy is maintained among all the

Table 2 Authentication and confidentiality analysis

	Confidentiality			Authentication		
Protocol	MN-FA	FA-HA	MN-HA	MN-FA	FA-HA	MN-HA
Protocol in [3]	None	None	Yes	None	None	MAC^S
Protocol in [4]	None	None	Yes	DS	DS	DS
Protocol in [5]	Yes	None	Yes	None	None	DS
Protocol in [6]	Yes	Yes	Yes	None	DS	SE
Time-invariant	Yes	Yes	Yes	None	MAC^S	MAC^D
Time-variant	Yes	Yes	Yes	None	MAC^D	MAC^D
CL-PKI	Yes	Yes	Yes	None	DS	MAC^D
ID-based [10]	Yes	Yes	Yes	None	IBS	MAC^D
Proposed	Yes	Yes	Yes	Hash	Hash	Hash

DS digital signature, MAC^S, MAC^D MAC using static and dynamic key respectively, *SE* symmetric encryption

entities in the proposed registration. Prior to the verification of the message sent between the parties of the communication, the communicants must be able to authenticate each other. The proposed registration gives the authentication between all the pairs and it is achieved with the hash functions; the hash value, $H(P_2)$, ensures the authentication on MN; $H(P_3)$ and $H(P_5)$ guarantee the authentication on FA for HA and MN; and the hash value $H(P_4)$ ensures the authentication on HA. Table 2 shows the authentication and confidentiality analysis of the existing protocols and the proposed protocol.

The timeliness of the message is an important issue because of the threat of message replays. To prevent replay attack, an arbitrary number called nonce is used between the communicants. In the proposed protocol, the replay attack prevention is provided by using the nonces MN_{Nonce}, HA_{Nonce}, and FA_{Nonce}. The user anonymity is also attained by using the temporary identity MN_{tid}, instead of its original identity. Table 3 shows the replay attack prevention and user anonymity analysis.

5 Formal Verification and Performance Evaluation

The security properties of the proposed protocol are validated using AVISPA [17]. The proposed sequence is specified by using high-level protocol specification language (HLPSL) which is a state machine and role-based descriptions [18]. Subsequently, these descriptions are translated into an equivalent intermediate format (IF). Then, it is given to the back-ends of AVISPA that implement automatic analysis techniques such as on-the-fly model checker (OFMC) and constraint-logic-based attack searcher (CL-AtSe) for detecting any attacks.

In our protocol simulation in AVISPA, we defined three roles in HLPSL for the entities MN, FA, and HA. The simulation starts with the state transition 0 and follows with the respective transitions from 1 to 8 for successful execution of the

Table 3 Replay attack prevention and user anonymity analysis

Protocol	In [3–5]	In [6]	TIV	TV	CL-PKI	In [10]	Proposed
Prevention of replay attack	None	Yes	Yes	Yes	Yes	Yes	Yes (Nonce)
User anonymity	None	Yes	Yes	Yes	Yes	Yes	Yes (Temp.id)

OFMC results:
SUMMARY
SAFE
DETAILS
BOUNDED_NUMBER_OF_SESSIONS
PROTOCOL
/root/avispa-1.1
/results/FBS_MobileIPFractal.if
GOAL
as_specified
BACKEND
OFMC
COMMENTS
STATISTICS
parseTime: 0.02s
searchTime: 0.13s
visitedNodes: 75nodes
depth: 7 plies

CL-AtSe results:
SUMMARY
SAFE
DETAILS
BOUNDED_NUMBER_OF_SESSIONS
TYPED_MODEL
PROTOCOL
/root/avispa-1.1
/results/FBS_MobileIFractalP.if
GOAL
As Specified
BACKEND
CL-AtSe
STATISTICS
Analysed : 67 states
Reachable : 13 states
Translation: 0.03 seconds
Computation: 0.82 seconds

Fig. 2 Back-end results of the simulated protocol in AVISPA

protocol. Then, the HLPSL has been translated into IF. It was tested automatically for attack tracing with OFMC and CL-AtSe results as shown in Fig. 2. No attacks were found and the security goals are verified for our proposed protocol.

The protocol includes the amount of signalling traffic in terms of message size for secure transmission of location update messages during registration. In the proposed protocol, the request message size between MN and FA is 78 bytes; the request message size between FA and HA is 160 bytes; the reply message size between HA and FA is 152 bytes; the reply message size is 45 bytes. The numerical values of the proposed protocol such as cryptography operation time on FA, HA, and MN are obtained from [19] and the system parameters from [20].

The total registration time of the proposed protocol including all the steps = 2.8311 + 1.013698 + 1.196192 + 6.45 = 11.4909 ms.

From Fig. 3, it can be observed that the processing time for request message between FA and MN of our proposed scheme is reduced when compared to other protocols. Some of the protocols show different processing time depending upon the security computations that are involved in the registration. The two protocols, time-variant and time-invariant, are varied with the processing time in terms of two different aspects of security and efficiency. The protocol in [6] offers high delay in processing request message between FA and HA since it requires the use of public

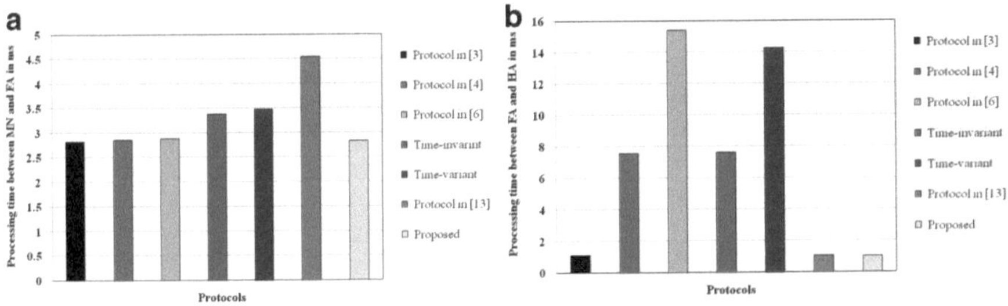

Fig. 3 Request message processing time between (**a**) MN-FA and (**b**) FA-HA

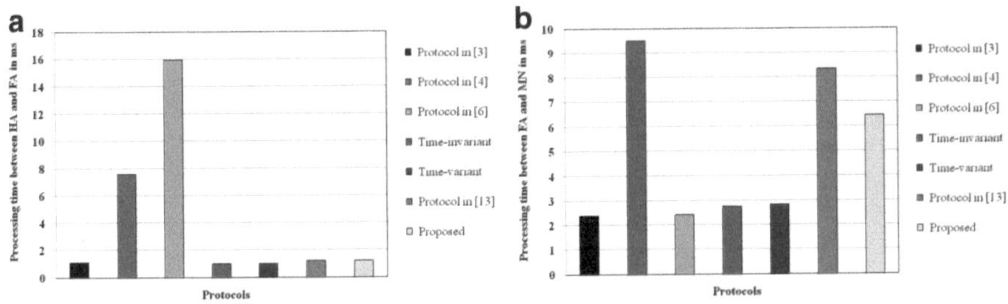

Fig. 4 Reply message processing time between (**a**) HA-FA and (**b**) FA-MN

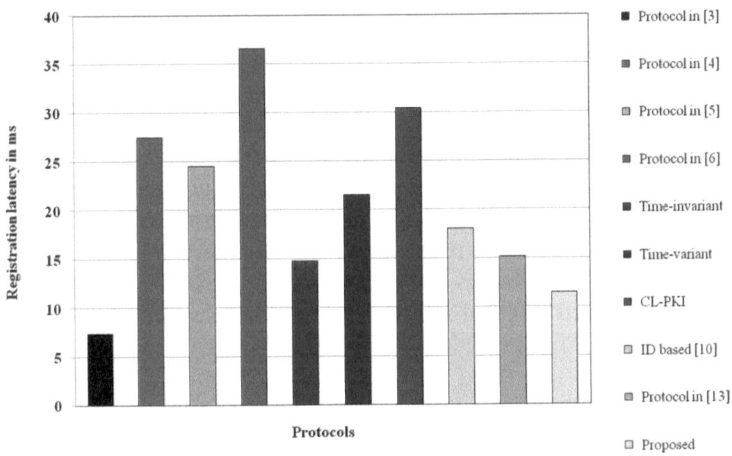

Fig. 5 Registration delay of the compared protocols

key cryptography to establish a session key. Subsequently, we can see that our protocol gives a shorter time for processing request message between FA and HA.

Figure 4a suggests that most of the protocols pay out almost the same processing time to accomplish reply message except protocols [4] and [13]. The reason is that there are more operations needed in validating the messages in HA. Figure 4b shows the high delay of the proposed one for processing reply in average, since it requires the verification on reply message between FA and MN. Next, we examine the total delay in completing the registration. The individual operations are summed and plotted in Fig. 5. It is observed that the base protocol is provided with lesser registration time but it suffers from security threats as we have discussed in the related works. Clearly, the graph shows that the performance of the proposed one stands out among other protocols with balanced processing time.

6 Conclusion

The location update process must be designed with the balancing effort on security and efficiency in terms of registration time; hence, we dealt with a balanced framework for a secure and efficient mobile IP registration, which provides better security and significant reduction in registration delay. In this chapter, the location update protocol is proposed based on fractal scheme to minimize the registration latency and for security attributes such as authentication, confidentiality, replay attack prevention, and user anonymity. The performance analysis and the numerical findings have also been discussed for security and performance considerations.

References

1. Perkins, C.E.: IP encapsulation within IP. RFC 2003, pp. 1–18. (1996)
2. Senthil Kumar, M., Valarmathi, M.L.: Mobile IP registration protocols: a survey. Int. J. Comput. Appl. **51**(17), 24–34 (2012)
3. Perkins, C.E.: Mobile IP. IEEE Communication Magazine, pp. 84–99. (1997)
4. Jacobs, S.: Mobile IP Public Key based Authentication. http://www.search/ietf.org/Internetdrafts/draftjacobs-mobileip-pkiauth-01.txt (1999)
5. Sufatrio, Lam, K.Y.: Mobile-IP registration protocol: a security attack and new secure minimal public-key based authentication. In: International Symposium on Parallel Architectures. (1999)
6. Yang, C.Y., Shiu, C.Y.: A secure mobile IP registration protocol. Int. J. Netw. Secur. **1**(1), 38–45 (2005)
7. Dang, L., Kou, W., Zhang, J., Cao, X., Liu, J.: Improvement of Mobile IP registration using self-certified public keys. IEEE Trans. Mobile Comput. (2007)
8. Dang, L., Kou, W., Dang, N., Li, H., Zhan, B.: Mobile IP registration in certificateless public Key infrastructure. IET Inf. Secur. **1**(4), 167–173 (2007)
9. Manjun, Zhang, Changxing, Pei, Lanjun Dang: Efficient mobile IP registration in certificateless signature. In: IEEE 24th International Conference on Advanced Information Networking and Applications Workshops, pp. 363–366. (2010)
10. Dang, L., Kou, W., Li, H., Zhang, J., Cao, X., Zhao, B.: Efficient ID-based registration protocol featured with user anonymity in mobile IP networks. IEEE Trans. Wireless Commun. **9**(2), 594–604 (2010)
11. He, Liu, Mangui, Liang: Efficient Identity-based hierarchical access authentication protocol for mobile network. Secur. Commun. Netw. **6**(12), 1509–1521 (2013)

12. He, Liu, Mangui Liang: Privacy-preserving registration protocol for mobile network. Int. J. Commun. Syst. (2012)
13. Senthil Kumar, M., Valarmathi, M.L., Ramprasath, G.: A secure and efficient registration for IP mobility. In: Proceedings of the First International Conference on Security of Internet of Things, ACM, Kollam, India, pp. 210–215. (2012)
14. Senthil Kumar, M., Valarmathi, M.L.: An efficacious and secure registration for internet protocol mobility. Defence Sci. J. **63**(5), 502–507 (2013)
15. Alia, M.A., Samsudin, A.B.: New key exchange protocol based on mandelbrot and julia fractal sets. Int. J. Comput. Sci. Netw. Secur. **7**(2), (2007)
16. Patrzalek, E.: Fractals: Useful Beauty General Introduction to Fractal Geometry. Stan Ackermans Institute, Centre for User-System Interaction, Eindhoven University of Technology (2006)
17. Armando, A., Basin, D., Cuellar, J., Rusinowitch, M., Vigano, L.: Automated Validation of Internet Security Protocols and Applications. ERCIM News (2006)
18. AVISPA v1.1 User Manual. http://www.avispa-project.org
19. Dai, W.: Crypto++ 5.2.1 Benchmarks. http://www.eskimo.com/~weidai/benchmarks.html (2004)
20. Argyroudis, P.G., Verma, R., Tewari, H., Mahony, D.O.: Performance analysis of cryptographic protocols on handheld devices. http://www.cs.tcd.ie/publications/techreports/reports.03/TCD-CS-2003-46.pdf (2003)

A Novel-Based Multi-agent Brokering Approach for Job Scheduling in a Cloud Environment

G. Nithya, R. Engels, H. Rahul Das, and G. Jayapratha

Abstract With the recent developments in the field of science and technology, the capabilities of handling complex problems have increased because of the maximum usage and management of computing power. Real resources are allocated are of reality, but the difficult part lies in proper identification and accumulation of resources which are required for solving complex problems. However, to get rid of this issue, the current trend is to use cloud computing effectively by resource sharing. The final objective is to facilitate optimum utilization and computing by aggregating idle network and processing resources like CPU cycle and storage spaces. Therefore, an effective measure has to be implemented so as to meet the job requirements by identifying appropriate service providers for successful execution. The allocation and scheduling of jobs should solve various problems and promote optimum utilization of resources. The key objective of this project is to identify and solve various problems mentioned above with the help of the multi-agent brokering approach and the jumper firefly algorithm (JFA). The multi-agent brokering

G. Nithya (✉)
Department of Information Technology, Sri Krishna College of Engineering and Technology, Coimbatore, Tamil Nadu, India
e-mail: nithyag@skcet.ac.in

R. Engels
Department of Computer Science and Engineering, PSG College of Technology, Coimbatore, Tamil Nadu, India
e-mail: rengels@outlook.com

H.R. Das
PSG Institute of Management, Coimbatore, Tamil Nadu, India
e-mail: das045@gmail.com

G. Jayapratha
Department of ECE, Bannari Amman Institute of Technology, Sathyamangalam, Tamil Nadu, India
e-mail: jayaprathag93@gmail.com

E.B. Rajsingh et al. (eds.), *Informatics and Communication Technologies for Societal Development*, DOI 10.1007/978-81-322-1916-3_8, © Springer India 2015

approach helps in the selection of various service providers in a cloud environment, and jumper firefly helps in reducing the make span time by its status table by recording the behavior of each firefly in detail. The proposed algorithm makes the weaker ones to jump to a new position so as to attain high probability. Hence, this helps to attain a better performance in finding an optimal solution to various complex issues. From various experimental angles, the jumper firefly mechanism is considered more efficient in terms of the make span time than the standard firefly algorithm (SFA) or any other methods.

Keywords Cloud computing • Multi-agent brokering approach • Genetic algorithm • Heuristic methods • Jumper firefly algorithm

1 Introduction

The job processing system in a large-scale environment helps in the prompt servicing of jobs and optimum utilization of resources. The key challenge is to schedule incoming jobs by satisfying or meeting the above requirements by using various job scheduling schemes. In the middle of 1990s, cloud computing was used to satisfy the rising demands such as bandwidth, storage, speed, capacity and resources [12]. The greatest challenges in a cloud environment are resource management and optimum utilization as well as the various management policies [1] practiced.

In the current scenario, cloud computing plays a vital role in the reduction of computational complexity and also provides many services using virtualized resources [20]. Efficient resource sharing is done by the end users within a reasonable amount of time. There are two types of cloud: one is the public cloud which is sold to anyone via Internet connection. The recent real time applications are Amazon Elastic Compute Cloud (EC2) and Google App Engine. The other is the private cloud, which is the proprietary network or a data center that supplies services to a limited number of clients [15].

Several information technology services are available, such as Software as a Service (SAAS), which enable the users to run applications remotely. Information as a Service (IAAS) is an application which is located via cloud. Platform as a Service (PAAS) offers solution stack, programming language, and operating system. The well known cloud is the Amazon Elastic Compute Cloud (EC2), which is IAAS type of cloud. EC2 clients are charged based on CPU per hour virtual machine. It has a capability to create multiple instances as fast as possible. It has a capability to create multiple instances as fast as possible, at the same time clusters can form base virtual machine. However, it is very much necessary to apply an approach which guarantees the distribution or proper allocation of workloads across thousands of users' queries per second, e.g., the job scheduling approach [14]. In a cloud environment, commodity servers also perform the job scheduling process.

Scheduling jobs on a cloud environment is said to be an NP-complete problem [10]. The standard firefly algorithm (SFA) is a meta-heuristics technique which is processed by the flashing behavior of fireflies. It is also known as a population-based technique as it helps to find the global optimal solution which investigates the foraging behavior of fireflies [22].

Novel method-based firefly algorithms are introduced as they help in maximizing the standard firefly algorithm sharing information with the help of a status table. It is also required to investigate the situation of fireflies in terms of finding the proper solution and identifies the most inefficient agent by utilizing the jump option [23]. This proposed algorithm aims to produce optimal schedule of jobs by assigning end user tasks to the relevant data centers and also minimize the waiting time, thereby outperforming other optimization methods in terms of search space and convergence. Agent technology can be effectively used for finding suitable job processing units in a distributed shared resource environment using the multi-agent approach. Therefore, we can ensure that the agents specialize in the identification of a service provider for scheduling jobs.

The three key agents considered for this project start with a job requestor, an agent that is a software program to find appropriate providers with the appropriate computing resources for executing the jobs, through a mapper agent [16]. Secondly, the job scheduler agent provides job scheduling, for requests from a mapper agent, and maintains information in service providers for better mapping. The job mapper agent connects agents and facilitates the matching of job requests from users to appropriate computing resources of providers. It may also be used to cope up with unexpected events.

The proposed approach is scalable by using multiple broker agents. The initial motivation for the work presented in this chapter is to identify the job schedule plan [17]. It also assigns end user tasks to the relevant data centers in an optimal fashion. Using multi-agents provides efficient job scheduling in cloud environment [24].

The ideas mentioned below can be implemented:

- Reducing the make span or waiting time in a cloud environment
- Designing user agents and provider agents that help in handling job requests and processing unit information
- Using shared and combined resources to produce an optimal schedule by minimizing the average waiting time and waiting time

The remainder of the chapter is organized as follows: Sect. 2 describes the related work; Sect. 3 explains the current methodology used and focuses on the proposed approach and its parameters used; Sect. 4 describes the standard firefly algorithm; Sect. 5 provides an understanding of the proposed jumper firefly algorithm (JFA); Sect. 6 presents the experimental results; and Sect. 7 concludes the chapter with future work and finally the references of this chapter are also provided.

2 Related Work

2.1 Agent Team Approach

According to this approach [2], the agent team is taken into consideration in the initial stage as it takes care after the work about scheduling, but the negative part lies when the agent's leader fails as it cannot proceed. The agent can differentiate in performing work and more specifically the execution becomes difficult if the agent fails.

2.2 A Multi-agent Brokering Protocol

There are three types of agents that are used in this multi-agent protocol. They are the requestor agent, mapper agent, and scheduler agent [11]. Each mapper agent connects requestor agents to scheduler agents using the connection algorithm. The connection algorithm consists of four stages: selection, evaluation, filtering, and recommendation. In the recommendation stage, there are two kinds of recommendation approaches: one is the circular approach and the other one is the multicast approach [13]. These approaches are used for making recommendations to the requestor agents that failed to be matched with the mapper agents.

In [3], the author proposed a design of a workload generator to perform alternative configuration with a real workload. That kind of design helps the system to perform job scheduling, resource sharing, management, and workload with its workload parameter. In [4], the author proposed a load balancing framework for cluster data storage in a cloud environment. This helps in balancing load and minimizes the computational complexity [20]. This framework is demonstrated using NetApp Data ONTAP UX system and the clustered NFS server system, which minimize the reconfiguration cost in nature [18].

There are several greedy techniques that are integrated to enhance in a single step-in which finds the optimal solution in a faster way. For example, many algorithms simulated annealing and genetic which may struck at local optima. The ACO is an optimization technique which is inspired by real ants to describe the shortest path to reach the location prescribed. The ants move in search of food and back to their nest while dropping pheromones on the path to identify it and also encourage other ants. The process continues with more deposits of pheromone and check the path which is no longer used, since the pheromone will start to evaporate [22]. Particle swarm optimization (PSO) is one of the swarm intelligent optimization techniques inspired by some social behavior of flocking birds or schooling fish [5]. This records the best solution which is already discussed. It also has a capability to update the position vector and velocity by global best (G BEST) and local best (P BEST) positions. The firefly algorithm is termed as a stronger meta-heuristics algorithm than any other optimization methods. It also solves problems like selection, scheduling, searching, and traveling salesman with its optimum waiting time, but the status table investigates the situation of fireflies which is not recorded [19].

3 Proposed System

Effective methods of identifying service providers based on all the needs of jobs for successful execution are required for the proposed system. The assigning and scheduling of the jobs should result in efficient problem solving and should promote optimum utilization of resources. This project proposes to address the above problem using the multi-agent brokering approach for the identification of service providers in a cloud environment. The multi-agent approach helps to ensure that the agents are able to specialize in the identification of a service provider for scheduling jobs [22]. For achieving this, proper representation of provider capabilities and job needs is considered to be essential [17, 21].

This chapter aims at minimizing the make span or waiting time, whereas the make span time is considered as one of the critical factors in a cloud environment. The effectiveness and efficiency of each algorithm depends on the make span time. Scheduling the longest job on the fastest resource (LJFR) rule minimizes the make span time. It is a fact that by minimizing the make span time, flow time is maximized, and vice versa [5]. This chapter assumes that resources employed are by first come, first served (FCFS) policy for executing received jobs.

It can be explained with the help of a deployment diagram and an activity diagram which are represented in Figs. 1 and 2.

The following results are achieved using proposed approach:

- Scalability
- Identifying efficient service providers by this approach
- Reducing make span/waiting time
- Taking multiple factors for jobs into consideration
- Extending jade technologies
- Job schedule plan identification with the help of agent technology
- Identifying weak agents
- Tracking status record with status table
- Identifying the behavior of firefly in each move

4 Problem Statement and Formulation

The primary focus of job scheduling is to allocate jobs to data centers present in the cloud. This problem can be formulated as follows: Jobs = $\{J_1, J_2, \ldots, J_i, \ldots, J_n\}$, a set of "n" jobs to be scheduled. $Job_i Tasks = \{JTask_{i1}, JTask_{i2}, \ldots, JTask_{im}\}$, as a set of

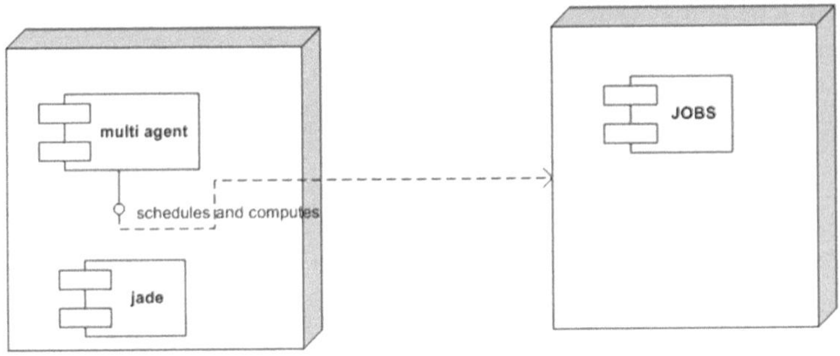

Fig. 1 Deployment diagram

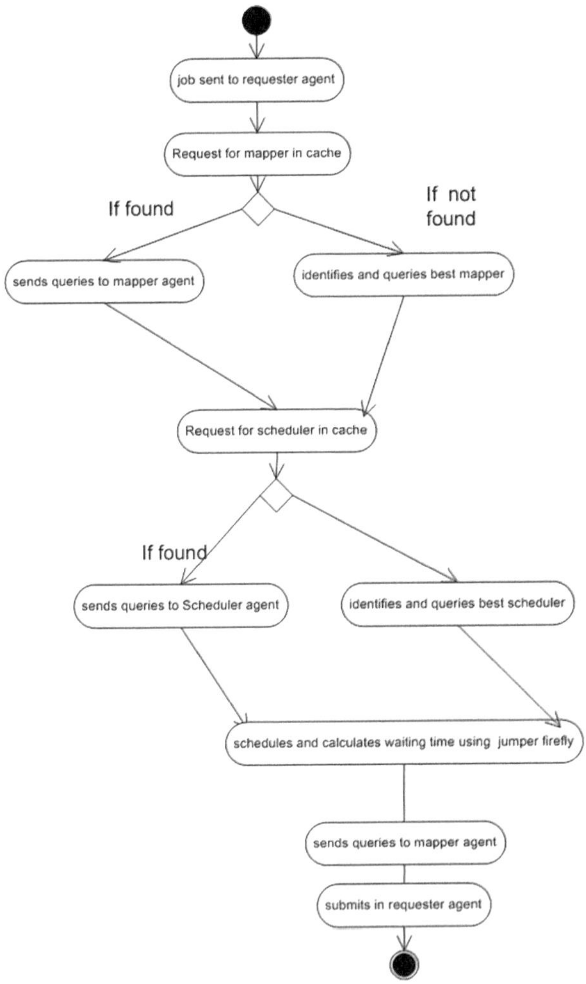

Fig. 2 Activity diagram

"m" task partitions of `Job_i` among "m" cloud data centers (DCs) in order to be executed.

DC_j ensures the execution of their tasks as follows:

$DC_jTasks = \{JTask_{aj}, JTask_{bj}, \ldots, JTask_{rj}\}$

Total execution time of all job tasks DCj would be

Make span $(DC_jTasks) = Max(JTask_{kj}.StartTime + JTask_{kj}.ExeTime)$. In cloud computing the job scheduling is defined as follows:

$DCTasks = \{DC_1Tasks, DC_2Tasks, \ldots, DC_mTasks\}$

$DC_jTasks = \{JTask_{aj}, JTask_{bj}, \ldots, JTask_{rj}\}$ for r tasks < n which reduces make span (DC_jTasks)

A fitness function is defined as follows (used to calculate the above waiting time):

Fitness (DCTasks) = Σ (Fitness (JTaskij, DCj)) where $(1 \leq j \leq m)$

Fitness $(JTask_{ij}, DC_j) = JTask_{ij}.TimeToExe$ where $JTask_{ij}.TimeToExe$ are the execution time of task of job "i" that needs to run in DC_j

5 Standard Firefly Algorithm (SFA)

Firefly algorithm was introduced in 2008 by Mr. Yang, and it was inspired by the flashing behavior of the population-based technique to find the global optimal solution which helps in the investigation of the foraging behavior of fireflies. The key function of firefly's flash is to attract other fireflies. The flashing signal by fireflies is to attract mating partners [16].

The three key rules governing the behavior of fireflies, proposed by Mr. Yang, are as follows:

- All the fireflies are unisexual.
- The attractiveness is proportional to their brightness, and for any two fireflies, the one which has lesser brightness will be attracted to the brighter one. However, the brightness attenuates over distance.
- The brightness of the fireflies indicates the quality of solutions is determined by fitness function which is defined by considering the problem [6, 7].

6 The Proposed Jumper Firefly Optimization Algorithm for Job Scheduling

A rethinking can be made on the agent goals and their population, so that it becomes very clear to state that the quality is determined by individual members in a larger population. Hence we have proposed to implement the standard firefly algorithm to fetch the most suitable one by modifying or improving the society of the fireflies. The jumper firefly algorithm is described below.

Status tables are used to record and monitor all details of the firefly's behavior, with the help of the table given below. This table includes shared data of fireflies and decision-making and modifying processes of weak agents in the standard firefly algorithm. Decision-making process is involved to execute search agents to realize whether any agent needs to use the jump option or not. The status table is shown in Table 1.

- Fitness indicates the quality of solution found by fireflies at the *i*th stage.
- Worst indicates the number of worst solutions during searching phases.
- Finally, qualification indicates the fitness of each firefly from the beginning of search process.

6.1 *Jumper Firefly Algorithm*

```
Input: F(X), X = (X1, X2, X3,...,Xn)
m, γ, α // user-defined constant
Output:. I = F(X)
   for i =1: m
    Xi = initial_solution()
     end i
      Create Status Table
     While (termination requirements are not met) do
         If (any firefly is in hazard)
         Put firefly in new position stochastically
            Update Status Table
            end if
             min ← arg min i←{1,...,m} (F(Xi))
               for i = 1: m do
                 for j = 1 : m do
                  if F(Xi) < F(Xj) then
                   di,j = Distance(Xi,Xj)
```

(continued)

Table 1 Status table

Firefly	1	2	3	N
Situation				
Qualification				
Worst				

```
            β ← attractiveness(I0,γ,di,j)
        Xi ← (1- β)Xi + βXj + a(random()-1/2)
      Update Status Table
     end if
    end j
   end i
 Xmin ← X min + a(random() - 1/2)
 end.
```

6.1.1 Solution Representation

Each population contains "N" individuals (solutions) where each one of them is represented by a set of data centers.

Each data center carries out a set of job tasks as follows:

DCTasks = {DC1Tasks, DC2Tasks... DCmTasks}: set of data centers

Each data center contains a set of affected job tasks as follows:

DC_jTasks = {$JTask_{aj}$, $JTask_{bj}$... $JTask_{rj}$}

The JFA initialization focuses on the generation of the first population in which "N" individuals are randomly selected. For example, the following individuals are selected:

DCTasks = {DC1Tasks,DC2Tasks,...,DCmTasks} = {<JTaska1,JTaskb1,..., JTaskr1>,<JTaska'2,JTaskb'2,...,JTaskr'2>,...,<JTaska"m,JTaskb"m,..., JTaskr"m>}

Whereas R = {r_1, r_2... $r_{,m}$} are m cloud resources and J = {j_1, j_2,... J_n} are n-independent client jobs, t_{ij} represents the processing time of job j on resource i. Each task is characterized by its own execution time (JTaskij.TimeToExe).

6.1.2 Initial Population

This algorithm generates the initial population of feasible candidates of size *N* [8, 9]

$$x_{ij}{}^{0} = x_{min} + (x_{max} - x_{min}) + u(0, 1)$$

where $x_{min} = -0.4$, $x_{max} = 4.0$, and $u(0, 1)$ random values and $0 \leq u(1, 0) \leq 1$.

The fitness function and attractiveness are to determine the quality of a given candidate solution.

A fitness function is defined as follows

Fitness (DCTasks) = Σ (Fitness ($JTask_{ij}$, DC_j)) where $(1 \leq j \leq m)$ and Fitness ($JTask_{ij}$, DCj) = JTaskij.TimeToExe

where *JTaskij.TimeToExe* is the execution time of task of job "*i*" that needs to run in *DCj*.

6.1.3 Evaluate the New Solution and Update the Light Intensity

6.1.4 Rank the Fireflies and Find the Current Best Solution and Iterate the Parameter

6.1.5 Update the Status Table

New situation for fireflies means the re-initialization and rearranging into new location. As the re-initialization must preferably be done randomly, the algorithm assigns the weakest firefly stochastically in the new location:

Situation (firefly (m)) = new situation of firefly (m)
Fitness (firefly (m)) = fitness of newly found solution
Worst (firefly (m)) =0
Qualification (firefly (m)) = AVE (Qualification)

To perform a local search to reach the best individual from the original individual, the task of one data center is randomly selected with another task in the nearest data center as follows:

$DC_iTasks = \{<JTask_{ai}, JTask_{bi}, JTask_{ci}, JTask_{di}, \ldots, JTask_{ri}>\}$

Here we assume that DC_kTasks is the nearest data center to DC_iTasks, where $DC_kTasks = \{<JTask_{ak}, JTask_{Yk}, JTask_{ck}, JTask_{dk}, \ldots, JTask_{rk}>\}$

Then, the neighboring individual will be $DC_iTasks=\{<JTask_{ai}, JTask_{Yi}, JTask_{ci}, JTask_{di}, \ldots, JTask_{ri}>$

Table 2 Calculation of execution time of single task into data center

Dc_j	1	2	3	4	5	6	7	8
Exe time	10	11	12	13	12	12	13	11
Dc_j	9	10	11	12	13	14	15	16
Exe time	11	11	9	14	15	8	16	13

Fig. 3 Firefly representation and mapping

Resources 1 2 3 4
Jobs 1 2 3 4 5 6 7

Dimension(Jobs) 1 2 3 4 5 6 7
Firefly position $X_{1,1}{}^K$ $X_{1,2}{}^K$....... $X_{1,7}{}^K$
Value (resources) 2 4 1 4 2 5 1

7 Experimental Study

7.1 Experimental Settings

With the help of 16 data centers, a series of tests are carried out in the virtual environment in order to evaluate the performance and the effectiveness of JFA to schedule tasks between data centers in the cloud.

Jobs are classified into different tasks, which are considered more relevant to the existent data centers. Each task is identified by its own identifier and the execution time. In order to deal with JFA, the parameters are chosen as follows: the population size is equal to 10 individuals; here iteration is fixed as 30 and maximum as 150. Table 2 indicates the execution time (measured in milliseconds) of one task by a data center.

Simple firefly mapping with 7 client jobs to 5 cloud resources is shown in Fig. 3.

7.2 Experimental Results

After the SFA and JFA test executions by following the same parameter settings according to the cloud, the best solutions ascertained (tasks repartition) in terms of waiting time are listed in Table 3.

In order to calculate the fitness, the below mentioned fitness function is applied: Fitness (DCTasks) $-\Sigma$ (Fitness ($JTask_{ij}$, DC_j)) = Fitness ($JTask_{11}$, DC_1) + Fitness ($JTask_{1,5}$, DC_5) + ... + Fitness($JTask_{5,19}$, DC_{19}) + Fitness($JTask_{7,16}$, DC_{16})where Fitness ($JTask_{ij}$, DC_j) = $JTask_{ij}$.TimeToExe.

For SFA we have Fitness (DCTasks) = (30) + (50) + (60) + (71) + (71) = 282 ms. For JFA, we have Fitness (DCTasks) = (30) + (40) + (50) + (60) + (68) = 248 ms. Table 3 shows that the best fitness is obtained by JFA (248 ms) compared with the fitness given by SFA (282 ms). It also shows the best execution time of the overall tasks of every single job (J1...J7) performed by the different data centers

Table 3 Task repartition between data centers

Algorithm	Task repartition between data centers
SFA	*Job1Tasks* = {*JTask1 1, JTask1 4, JTask1 6, JTask1 7, JTask1 5*}
	Job2Tasks = {*JTask2 1, JTask2 6, JTask2 7, JTask2 5, JTask2 9*}
	Job3Tasks = {*JTask3 1, JTask3 6, JTask3 7, JTask3 12, JTask3 13, JTask3 14*}
	Job4Tasks = {*JTask41, JTask4 4, JTask4 6, JTask4 7, JTask4 8*}
	Job5Tasks = {*JTask51, JTask5 6, JTask5 7, JTask5 8, JTask5 9*}
	Job6Tasks = {*JTask6 1, JTask6 6, JTask6 7, JTask6 12, JTask6 13, JTask6 14*}
	Job7Tasks = {*JTas7 1, JTask7 4, JTask7,6, JTask7, 7, JTask7 8*}
JFA	*Job1Tasks* = {*JTask1 1, JTask1 4, JTask1 6, JTask1 7, JTask1 5*}
	Job2Tasks = {*JTask2 1, JTask2 6, JTask2 7, JTask2 5, JTask2 9*}
	Job3Tasks = {*JTask3 1, JTask3 6, JTask3 7, JTask3 12, JTask3 13, JTask3 14*}
	Job4Tasks = {*JTask41, JTask4 4, JTask4 6, JTask4 7, JTask4 8*}
	Job5Tasks = {*JTask51, JTask5 6, JTask5 7, JTask5 8, JTask5 9*}
	Job6Tasks = {*JTask6 1, JTask6 6, JTask6 7, JTask6 12, JTask6 13, JTask6 14*}
	Job7Tasks = {*JTas7 1, JTask7 4, JTask7,6, JTask7, 7, JTask7 8*}

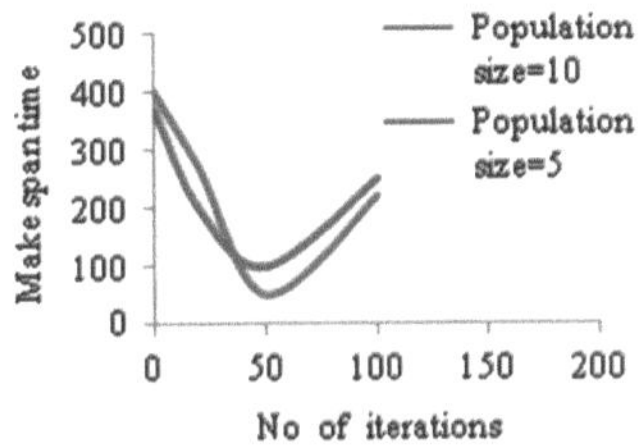

Fig. 4 Make span time for various population size

(DC1...DC16). Thereby these results point out the reliability and efficiency of JFA to solve job scheduling problem in the cloud by an optimal workload balancing.

The considered population sizes are 5 and 10. The waiting time is reported in Fig. 4. It can be seen in the population of size 10 which has a lesser waiting time than size of 5. The population size (number of fireflies) enhances the convergence and also indicates the population size that affects the searching space quality in the proposed mechanism.

8 Conclusion and Future Work

Therefore, by using the multi-agent approach, those jobs which are of multiple requirements are considered for scheduling in the cloud computing process. Here, a suitable service provider was identified and the waiting time for jobs was achieved by the jumper firefly algorithm, in which the status table takes a note of all the behavior about the fireflies. Here, the jump option was introduced as an added advantage which enables the fireflies to jump to a new situation in case of any hazardous situation. The result obtained from JFA proves the efficiency and the performance of the proposed algorithm in terms of execution time in a cloud computing environment.

This work is further extended using the following few practises: implementation of a service provider which enables tracking the status of change in time, identifying the factors affecting the performance of the system, and also benchmarking the proposed system with existing job scheduling techniques.

References

1. Huang, Y., Bessis, N., Norrington, P., Kuonen, P., Hirsbrunner, B.: Exploring decentralized dynamic scheduling for clouds and clouds using the community aware scheduling algorithm. Future Gener. Comput. Syst. **29**(1), 402–415, Jan 2013
2. Amazon Elastic Compute Cloud. http://aws.amazon.com/ec2/. Accessed 25 Oct 2011
3. Dominiak, M., Ganzha, M., Paprzycki, M.: Selecting cloud agent-team to execute user-job-initial solution. In: Proceedings of the Conference on Complex Intelligent and Software Intensive Systems, pp. 249–256. IEEE CS Press, Los Alamitos (2007)
4. Sim, K.M., Chan, R.: A brokering protocol for agent-based e-commerce. IEEE Trans. Syst. Man Cybern. Part C Appl. Rev. **30**(4), 474–484 (2000)
5. Sim, K.M.: Guest editorial: agent-based Cloud computing. Appl. Intell. **25**(2), 127–129 (2006)
6. Singh, K., Sonia: Optimized performance task scheduling algorithm for Cloud computing. Am. Int. J. Res. Sci. Technol. Eng. Math. **14**(2), 186–189 (2010)
7. Li, C., Li, L.: Multi economic agent interaction for optimizing the aggregate utility of Cloud users in computational cloud. Appl. Intell. **25**(2), 147–158 (2006)
8. Kang, J., Sim, K.M.: A brokering protocol for agent-based Cloud resource discovery. Commun. Comput. Inf. Sci. **63**(1), 33–40 (2009)
9. Naumenko, A., Nikitin, S., Terziyan, V.: Service matching in agent systems. Appl. Intell. **25** (2), 223–237 (2006)
10. Galstyan, A., Czajkowski, K., Lerman, K.: Resource allocation in the cloud with learning agents. J. Cloud Comput. **3**, 91–100 (2005)
11. Yosif, A.: Scheduling jobs in cloud environment using firefly algorithm. Int. J. Comput. Sci. Eng. **6**(1), 79–85 (2013)
12. Ardaiz, O., Artigas, P., Eymann, T., Freitag, F., Navarro, L., Reinicke, M.: The Catallaxy approach for decentralized economic based allocation in Cloud resource and service markets. Appl. Intell. **25**(2), 131–145 (2006)
13. Ganapathi, A., Chen, Y., Fox, A., Katz, R., Patterson, D.: Statistics-driven workload modeling for the cloud. University of California, Berkeley, Technical Report (2009)
14. Choi, S., Baik, M., Gil, J., Jung, S., Hwang, C.: Adaptive group scheduling mechanism using mobile agents in peer-to-peer Cloud computing environment. Appl. Intell. **25**(2), 199–221 (2006)
15. Dragoni, N., Gaspari, M., Guidi, D.: An infrastructure to support cooperation of knowledge-level agents on the semantic Cloud. Appl. Intell. **25**(2), 159–180 (2006)
16. Kunkle, D., Schindler, J.: A Load Balancing Framework for Clustered Storage. HiPC 2008, Lecture Notes in Computer Science, vol. 5374, pp. 57–72 (2008)
17. Fukuda, M., Kashiwagi, K., Kobayashi, S.: AgentTeamwork: coordinating Cloud-computing jobs with mobile agents. Appl. Intell. **25**(2), 181–198 (2006)
18. Goldberg, D.E.: Genetic Algorithms in Search, Optimization, and Machine Learning. Addison-Wesley, Boston (1989)
19. Jun Wu, Xin Xu, Pengcheng Zhang, Chunming Liu: A novel multi-agent reinforcement learning approach for job scheduling in Cloud computing. Future Gener. Comput. Syst.

27(5), 430–439 (2011). ISSN 0167 739x, 10.1016/j.future.2010.10.009. http://www.sciencedirect.com/science/article/pii/S0167739X10002025
20. Kang, Jaeyong, Sim, Kwang: A multiagent brokering protocol for supporting Cloud resource discovery. Appl. Intell. Springer, Dordrecht, 12 Apr 2012, pp. 1–16, ISSN 0924669X, 10.1007/s10489-012-0347-y. http://dx.doi.org./10.1007/s10489-012-0347-Y
21. Mathiyalagan, M.P., Suriya, S., Dr. Sivanandam, S.N.: Modified ant colony algorithm for Cloud scheduling. Int. J. Comput. Sci. Eng. (IJCSE), **02**(02), 132–139 (2010)
22. Naseer, A., Stergioulas, L.K.: Resource discovery in Clouds and other distributed environments: states of the art. Multi Agent Cloud Syst. **2**(2), 163–182 (2006)
23. Ruay- Shiung Chang, Chih-Yuan Lin, Chun-Fu Lin: An adaptive scoring job scheduling algorithm for cloud computing. Inf. Sci. **207**(10), 79/89, (2012). ISSN00200255, 10.1016/j.ins.2012.04.019. http://www.sciencedirect.com/science/article/pii/S0167739X1000202836
24. Tomassini, M.: Parallel and distributed evolutionary algorithms: a review. In: Miettinen, K., Makela, M., Periaux, J. (eds.) Evolutionary Algorithms in Engineering and Computer Science, pp. 113–133. Wiley, Chichester (1999)

A Comprehensive Survey on Data-Intensive Computing and MapReduce Paradigm in Cloud Computing Environments

Girish Neelakanta Iyer and Salaja Silas

Abstract With the emergence of cloud computing and other web technologies, data storage has become very easy and cheap. Hence, petabytes and zettabytes of data are in cloud storage on a daily basis through various portals such as YouTube and Facebook. This causes the need for big data analysis and data-intensive computing. In this paper, we perform a comprehensive survey on various data-intensive computing techniques and MapReduce paradigm mechanisms with cloud computing. We first provide data-intensive computing methodologies and then describe MapReduce algorithm for data-intensive computing on big data analysis.

Keywords MapReduce • Data-intensive computing • Hadoop

1 Introduction

Data-intensive computing is an emerging trend in the era of big data. Data-intensive computing is defined as the collection of data and programs or algorithms that are used to perform computations which provide higher levels of reliability, efficiency, availability, and scalability. In general, terabytes or petabytes of data are referred to as big data. Data-intensive computing is the data-parallel approach to process large volumes of data. Data-intensive computing is collecting, managing, analyzing, and understanding data at volumes and rates that push the frontier of current technologies.

Hadoop is an open-source cloud which is proposed by Yahoo. The framework for Hadoop is supported by Apache [1]. The main features of Hadoop are its

G.N. Iyer (✉)
Department of Information Technology, Karunya University, Coimbatore, TN, India
e-mail: girishniyer@gmail.com

S. Silas
Department of Information Technology, Karunya University, Coimbatore, TN, India
e-mail: blessingsalaja@gmail.com

E.B. Rajsingh et al. (eds.), *Informatics and Communication Technologies for Societal Development*, DOI 10.1007/978-81-322-1916-3_9, © Springer India 2015

scalability and reliability. Hadoop Distributed File System (HDFS) is the file system used by the Hadoop framework. Hadoop uses Java as the basic programming language, and it supports the programming methodology named MapReduce.

MapReduce is the programming model for big data processing which is inherently parallel. It is very simple and can be written in various languages such as Java, Python, Ruby, and C++. MapReduce divides applications into many small blocks of work. MapReduce has two functions, namely, map() and reduce(). The map function accepts input key/value pair and emits intermediate key/value pair. Map() procedure performs filtering and sorting (such as sorting students by first name into queues, one queue for each name). Another function is named Reduce() procedure which performs a summary operation (such as counting the number of students in each queue, yielding name frequencies). This paper aims to provide a survey in the area of both data-intensive computing and the MapReduce paradigm in Hadoop.

The remainder of the paper is organized as follows: In Sect. 1 we describe various data-intensive computing approaches. MapReduce in various fields are explained in Sect. 2. Section 3 concludes the paper.

2 Data-Intensive Computing Approaches

In this section, we briefly discuss about the approaches in data-intensive computing and also discuss about the methodologies used and the advantages and disadvantages of each approach.

C. Moretti et al. [2] described a new abstraction for data-intensive cloud computing, which mainly talks about solving all-pair problem for data-intensive applications. The methodology of the work is about the abstraction for data-intensive computing. This implements an optimized all-pair abstraction, which is easier to use than the underlying system and achieves performance orders of magnitude better than the obvious but naive approach and twice as fast as a hand-optimized conventional approach. The main advantage of the system includes optimized abstraction adjusted for grid and workload parameters to execute the workload efficiently and being more accessible to nonexpert users. The disadvantage of the system is that MapReduce cannot solve the all-pair problem efficiently and it limits the number of compute nodes. Also, the all-pair results can consume a tremendous amount of disk space.

R. Grossman et al. [3] proposed a method using sector and sphere, in which the main objective is to archive, analyze, and mine large distributed data sets using sector and sphere. This work aims at processing large data sets using sphere that are assumed to be broken up into several files. The work describes the programming paradigm supported by the Sphere compute cloud. Sector and Sphere are designed for analyzing large data sets using computer clusters connected with widearea high-performance networks. The main advantage of this Sector/Sphere is that it is approximately 2.4–2.6 times faster than Hadoop using the TeraSort benchmark supplied with Hadoop.

Y. Yu et al. [4] proposed that a high-level programming language for distributed data-parallel computing made data-parallel cluster computing as easy as writing programs for a single computer. It also gives design and implementation of the Dryad and DryadLINQ systems. It can be used with small, medium, and large clusters for nearly 3 years. A DryadLINQ program is a sequential program composed of LINQ expressions performing arbitrary side-effect-free operations on data sets and can be written and debugged using standard .NET development tools. The DryadLINQ system automatically and transparently translates the data-parallel portions of the program into a distributed execution plan which is passed to the Dryad execution platform. A DryadLINQ program is a sequential program composed of LINQ expressions. Dryad design is the support for dynamic optimization. The disadvantage of the system is it only supports for a specific set of application in .NET but also it is not as simple as MapReduce.

Jianwu Wang [5] proposed a new architecture for data-intensive applications in scientific work flow systems. The integration of Kepler and Hadoop is made in this work. This provides an easy-to-use architecture which helps the users to compose and execute MapReduce applications in Kepler scientific work flows. The characteristic features of scientific work flow management systems are very complementary to those of MapReduce. With this method, a Kepler graphical user interface is used, and the scientists can easily utilize MapReduce in their domain-specific problems and connect them with other tasks in a work flow. The advantage of the system is that the parallel execution performance can be achieved without bringing its complexity to users.

3 MapReduce Programming Model and the Hadoop Framework

This section discusses about the MapReduce programming model in various environments and the Hadoop framework.

J. Dean [6] proposed a MapReduce interface tailored towards a cluster-based computing environment and large data processing in a small amount of time. The processing of satellite imagery data is also explained in this method using MapReduce. Users specify the computation in terms of a map and a reduce function, and the underlying run time system automatically parallelizes the computation across large-scale clusters of machines, handles machine failures, and schedules inter-machine communication to make efficient use of the network and disks. Programmers find the system easy to use: more than 10,000 distinct MapReduce programs have been implemented internally at Google over the past 4 years, and an average of 100,000 MapReduce jobs are executed on Google clusters every day, processing a total of more than 20 petabytes of data per day.

R. Chen [7] proposed a method to iteratively process small chunks of data in multicore platforms. It allows the exploits of the multicore environment for data-parallel applications and optimizing resource usages using the technology called Tiled-MapReduce. The methodology used is with nested parallel algorithms. The prototype of Tiled-MapReduce is based on Phoenix, an already highly optimized MapReduce run time for shared-memory multiprocessors. The prototype, namely, Ostrich, runs on an Intel machine with 16 cores. Tiled-MapReduce uses the tiling strategy to partition a large MapReduce job into a number of small sub-jobs and handles the sub-jobs iteratively. The main disadvantage of this system is that it supports only for small-sized data compared to the Hadoop framework.

Ranger et al. [8] evaluated the suitability of the MapReduce model for multicore and multiprocessor systems. The methodology used is the Phoenix run time API using MapReduce. It also automatically manages thread creation, dynamic task scheduling, data partitioning, and fault tolerance across processor nodes. Phoenix with multicore and symmetric multiprocessor systems evaluates its performance potential and error recovery features. MapReduce is a promising model for scalable performance on shared-memory systems with simple parallel code. The major advantages of the system include buffer management fault recovery, concurrency and locality management, and being able to automatically handle key scheduling decisions. The disadvantage of the system is that it only suits for shared-memory system. M. M. Rafique [9] proposed a methodology that deals with MapReduce on Large-Scale Asymmetric Multicore Clusters. The objective of the paper is setting with large-memory head nodes using general-purpose processors and small memory and regulating data distribution and streaming in MapReduce. Histogram takes input as bitmap image and produces the frequency count of each RGB color composition in the image, and k-means takes a set of points in an N-dimensional space and groups them into a predefined number of clusters with an approximately equal number of points in each cluster. It utilizes limited memory available at compute nodes efficiently, and it has minimal impact on manager load. It also scales well with increasing number of compute nodes. The disadvantage is that it is not suited for a virtualized environment and also the scientific data processing is not possible.

S. Ibrahim [10] proposed a new system named cloudlet, which is to overcome the overhead of VM while benefiting of the other features of VM. The algorithm used is greedy algorithm. The execution flow is as follows. Cloudlet consists of three main phases: (1) It starts with executing the maps locally in the VM after exporting the data from the DFS into its virtual disk. The advantage of this system is that it combines the features of data-intensive applications and MapReduce and the easy management system. But the cloudlet has the disadvantage of being not practically proved.

Jin et al. [11] proposed the realization of MapReduce for .NET-based data centers and implemented a high-performance library for the Message Passing Interface which improves the performance. The architecture explanation of this paper is that the run time system is based on the master–slave architecture. The four major phases are divided into two tasks: map and reduce. It uses a merge/sort mechanism. The major advantage of the system is that the performance is better than Hadoop in the merge + reduce phase. It is also able to provide scalable performance in distributed environments. The major disadvantages of the system are that the sorting algorithm implemented by Hadoop is more efficient and it only gives good performance in windows machines (XP) practically.

C. Miceli et al. [12] proposed a system to create distributed applications in an infrastructure-independent way using SAGA (Simple API for Grid Applications) which is implemented using MapReduce. The two processes are master and worker processes, where the master process deals with coordinating the executed workers including the chunking of the data and assigning the input data to the workers of the map step, whereas the worker processes get assigned work from either the map or the reduce step. The major disadvantage is that the system scalability is not possible beyond 10 nodes.

A. Dou et al. [13] proposed a framework which provides a powerful software abstraction that hides many complexities from the application developer. The design implemented a mobile MapReduce framework targeted at any device which supports Python and network connectivity using Misco, a MapReduce Framework for Mobile Systems. The methodology used in this paper is a single map or reduce algorithm which is implemented in Python. Input data is submitted by the user; it is split into many pieces. In reducer, all the files belonging to the same partition but produced by different map tasks are merged together. The merged data file containing the map result from all the map tasks in a job is used as the input for a reduce partition. The user is to decide how many partitions the map data will be split into and thus how many reduce tasks will be created. A reduce input data along with a reduce operation forms a reduce task. The advantage of the system is an assisted living application for monitoring the activity of elderly people at home. This system's disadvantage is its applicability to mobile environments only.

Lizhe Wanga et al. [14] integrated a new file system named Gfarm file system with Hadoop framework and compared it with the HDFS file system. This method has high scalability, and there are no limitations in using the number of data nodes. It uses BBP algorithm for comparing the results. The time complexity is reduced in this method. It has high efficiency and easier distributed data processing. But the integration need for very high-end systems is not economical.

A comparison for MapReduce paradigm and data-intensive computing is provided based on the analysis of different features such as the basic mechanism, cost, efficiency, file system used, scalability, data nodes, and algorithms used in Table 1.

4 Conclusion and Future Enhancement

With the emergence of Hadoop and MapReduce paradigms, data-intensive computing received great attention in the recent past. In this paper, we discussed data-intensive computing and MapReduce paradigms in cloud computing environments.

Table 1 Comparison on data-intensive computing and MapReduce paradigm in Hadoop

Scheme	Basic mechanism	Scalability	Data nodes	Algorithms	Efficiency	Cost	File system
All pairs	Abstraction for DIC	Less chance	Limited	Active storage and demand paging	NO with MapReduce	High for disk space	HDFS, Chirp
Sector and Sphere	Distributed Sphere processing elements	High	Nil	Message passing	Only on wide-area high-performance networks	High	HDFS, GFS
High-level programming language	Dryad and DryadLINQ	High	Nil	Linear regression, expectation maximization	Only in .NET framework	Nil	NTFS
Data processing on large clusters	MapReduce	High	No limitations	Clustering algorithms	High	Low	HDFS
Optimizing resource usages	Tiled-MapReduce	Only on large-scale multicore systems	Nil	Nested parallel algorithms	High resource and memory efficiency	Nil	HDFS
Multicore and multiprocessor systems	MapReduce (Phoenix)	Good scalability on multicore platforms	No limitations	k-means, linear regression	Good	Nil	GFS, HDFS
Asymmetric multicore clusters	MapReduce	High	8 (can be high)	k-means, linear regression	Power efficiency	Low	NFS
MR on virtual machines	Cloudlet	High	7 (can be high)	Greedy algorithm	Moderate	Nil	AFS
.NET-based cloud computing	MapReduce	High	No limitations	MapReduce algorithm	Only for .NET-based data center	Nil	WinDFS, CIFS, or NTFS
DIC on cloud and grid	SAGA (simple API for grid applications)	High	Max 10 nodes	MapReduce algorithm	High(=1)	Nil	HDFS
	Misco	High				Nil	GFS

MR for mobile systems			No limitations	MapReduce algorithm using Python	Good in mobile environments		
MR across distributed data centers	G-Hadoop	High	No limitations	Word search, BBP algorithm	Very high	Nil	G-farm file system

We have provided a comparative study of various schemes and mechanisms used and of several features such as scalability, cost, and file system used.

The existing approaches have several drawbacks such as low efficiency, time complexity, and being difficult in practical realizations. Due to the inherent advantages of optimization algorithms such as ant colony optimization [15] or support vector machine [16], they could be used for data-intensive computing using the MapReduce paradigm. The use of such optimization algorithms in the MapReduce phase could help to reduce the time complexity and hence efficiency can be improved.

Acknowledgment The authors would like to acknowledge Dr. Ganesh Neelakanta Iyer for his valuable comments in completing this research work.

References

1. Shvachko, Hairong Kuang, Radia S., Chansler R.: The Hadoop Distributed file system. In: IEEE 26th Symposium on Mass Storage Systems and Technologies (MSST), (2010)
2. Moretti, C., Bulosan, J., Thain, D., Flynn, P.J.: All-pairs: an abstraction for data intensive cloud computing. In: IPDPS, April 2008
3. Grossman, R., Gu, Y.: Data mining using high performance data clouds: experimental studies using sector and sphere. In: Proceeding of the 14th ACM SIGKDD International Conference on Knowledge Discovery and Data Mining, KDD-08, ACM, New York, NY, USA (2008)
4. Yu, Y., Isard, M., Fetterly, D., Budiu, M., Erlingsson, U., Gunda, P.K., Currey, J.: Dryadlinq: a system for general-purpose distributed data-parallel computing using a high-level language. In: Proceedings of the 8th USENIX Conference on Operating Systems Design and Implementation, OSDI-08, USENIX Association, Berkeley, CA, USA (2008)
5. Jianwu Wang, Crawl, D., Altintas, I.: Kepler+Hadoop: a general architecture facilitating data-intensive applications in scientific workflow systems. In: WORKS 09, Portland Oregon, 15 Nov 2009
6. Dean, J., Ghemawat S.: Mapreduce: simplified data processing on large clusters. Commun. ACM **51**, (2008)
7. Chen, R., Chen, H., Zang, B.: Tiled-mapreduce: optimizing resource usages of data-parallel applications on multicore with tiling. In: Proceedings of the 19th International Conference on Parallel Architectures and Compilation Techniques, PACT10, ACM, New York, NY, USA (2010)
8. Ranger, C., Raghuraman, R., Penmetsa, A., Bradski, G.R., Kozyrakis, C.: Evaluating mapreduce for multi-core and multiprocessor systems. In: 13st International Conference on High-Performance Computer Architecture (2007)
9. Rafique, M.M., Rose, B., Butt, A.R., Nikolopoulos, D.S.: Supporting mapreduce on large-scale asymmetric multi-core clusters. SIGOPS Operat. Syst. Rev. **43**, 25–34 (2009)
10. Ibrahim, S., Jin, H., Cheng, B., Cao, H., Wu, S., Qi, L.: Cloudlet: towards mapreduce implementation on virtual machines. In: Proceedings of the 18th ACM International Symposium on High Performance Distributed Computing, HPDC 09, ACM, New York, NY, USA (2009).
11. Jin, C., Buyya, R.: Mapreduce programming model for.net-based cloud computing. In: Euro-Par. Lecture Notes in Computer Science, vol. 5704. Springer (2009)
12. Miceli, C., Miceli, M., Jha, S., Kaiser, H., Merzky, A.: Programming abstractions for data intensive computing on clouds and grids. In: Proceedings of the 2009 9th IEEE/ACM

International Symposium on Cluster Computing and the Grid, CCGRID09, IEEE Computer Society, Washington, DC, USA (2009)
13. Dou, A., Kalogeraki, V., Gunopulos, D., Mielikainen, T., Tuulos, V.H.: Misco: a mapreduce framework for mobile systems. In: Proceedings of the 3rd International Conference on PErvasive Technologies Related to Assistive Environments, PETRA10, ACM, New York, NY, USA (2010).
14. Lizhe Wanga, Jie Tao, Rajiv Ranjan, Holger Marten, Achim Streit, Jingying Chen, Dan Chena: G-Hadoop, MapReduce across distributed data centers for data intensive computing. Future Gener. Comput. Syst. **29**, (2013)
15. Xia Wei-Lei, Wang Li-Song: Research on and implementation of parallel ant colony algorithm based on MapReduce. Electron. Sci. Technol. **26**(2), 146 (2013)
16. Kiran, M., Kumar, A., Mukherjee, S., Ravi Prakash, G.: Program model for parallel support vector machine algorithm on hadoop cluster. Int. J. Comput. Sci. Iss. **10**(3), 317–325 (2013). no. 1, (May 2013)

Efficiently Scheduling Hadoop Cluster in Cloud Environment

D. Raghuram, N. Gayathri, and G. Sudha Sadasivam

Abstract Today, most of the real-time applications like bioinformatics and image processing involve processing of large amounts of unstructured data that requires fast, memory-consuming, and highly efficient resources. This problem has been resolved by the introduction of cloud, which is now the most favored option for big-data analytics. Hadoop, a framework for manipulating unstructured data, is used for this purpose. The nodes that form the Hadoop cluster are scheduled randomly in Amazon cloud. Since huge amounts of data need to be transferred among these nodes, the time taken to upload and process the data is quite high, thereby decreasing the performance. The further focus of service providers is on maximizing resource utilization and minimizing power consumption. This chapter aims at designing an energy-efficient scheduler for a cloud environment that will be suitable for the big-data applications. The working of the scheduler has been tested in OpenStack cloud environment.

Keywords Hadoop • OpenStack • Cloud computing • Energy efficiency

1 Introduction

Cloud computing [1] allows the user to gain access using the Internet, to off-site, increased computing capabilities (both storage and services) without having to maintain them. OpenStack [2] is a collection of open-source technologies that provide massively scalable open-source cloud computing software. It is an Infrastructure as a Service (IaaS) cloud computing project. Apache Hadoop [3] is an open-source Java software framework that supports massive data processing across a cluster of servers. Hadoop uses a programming model called MapReduce that

D. Raghuram • N. Gayathri (✉) • G. Sudha Sadasivam
Department of Computer Science and Engineering, PSG College of Technology, Coimbatore, TN, India
e-mail: cberaghuram@gmail.com; gayathri7492@gmail.com; sudhasadhasivam@yahoo.com

E.B. Rajsingh et al. (eds.), *Informatics and Communication Technologies for Societal Development*, DOI 10.1007/978-81-322-1916-3_10, © Springer India 2015

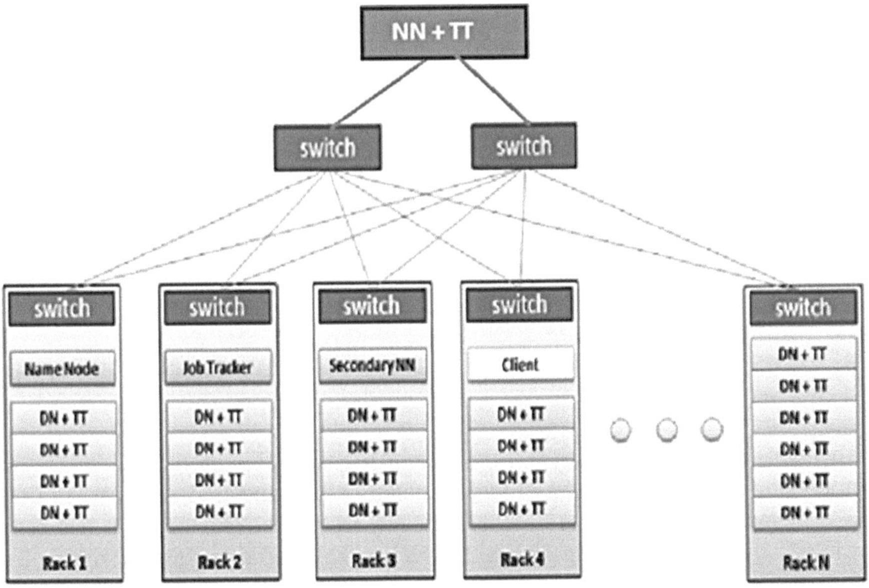

Fig. 1 Hadoop cluster

divides a large data set into many small fragments. Hadoop then combines the results from all of the nodes into a finished output.

1.1 Architecture and Working of Hadoop

Hadoop [3] is viewed as the perfect niche for big-data analytics as it can efficiently handle unstructured data. Hadoop solves this too by enabling distributed parallel processing of huge amounts of data. There are three kinds of machines available in a Hadoop cluster – master nodes, slave nodes, and client machines (Fig. 1).

1.2 OpenStack Components

OpenStack [2] consists of the following six components and scheduler framework is available in Nova:

- Horizon (Dashboard) which acts as the web front end to other OpenStack services
- Nova (Compute) which creates and terminates instances and stores the associated metadata in Glance (Image)
- Quantum (Network) which is responsible for the networking for Nova

- Cinder (Block Storage) which provides storage volumes for Nova
- Swift (Object Store) which stores the actual disk images
- Keystone (Identity) which authenticates access to all the above services

The rest of this chapter is structured as follows: Section 2 discusses the related works. Section 3 presents scheduling in OpenStack. Section 4 proposes approach. In Sects. 5 and 6, experiments are conducted and discussed. Section 7 gives the conclusions to this chapter.

2 Related Work

Traditionally algorithms implemented in cloud schedule the cloud nodes based on their QoS values, resource utilization, and reliability, without considering the communication performance between cloud nodes. However, such kind of node relationship is very important for the communication-intensive cloud applications, which have a lot of communications between the selected cloud nodes. Novel clustering-based method [4] selects optimal cloud nodes for deploying communication-intensive applications to the cloud environment. Dynamic resource allocation model [5] considers the location of physical machines and resource utilization of resources. If distance between nodes is far, the performance of the VM decreases as delay in communication increases. So schedulers must consider distance between nodes in data center.

Energy management in cloud computing has become a challenge as the number of data center is increasing. Beloglazov et al. [6] have studied and classified different ways to achieve power and energy efficiency in computing systems. The recent developments have been discussed, and techniques for achieving energy efficiency are categorized at hardware, operating system, virtualization, and data center levels. The authors [7] suggested that task consolidation is an effective technique to increase resource utilization. Also Beloglazov et al. have proposed an energy-aware task consolidation (ETC) [7] technique to increase resource utilization that in turn minimizes energy consumption. Energy-aware resource allocation heuristics [8] considers QoS expectations and power usage characteristics of the devices.

In IaaS, the instances are provided as VM based on SLA. Most IaaS does not consider the type of application loaded on VM. Consider a cloud provider, Amazon; they provide infrastructures as storage in S3, whereas computation is done in EC2 nodes. Hadoop works on the principle of maintaining computation and storage nodes. When Hadoop is loaded on instances provided by Amazon, the performance of Hadoop is reduced as data need to be transferred frequently between S3 and EC2. The authors [9] have studied the effectiveness of Hadoop cluster on AWS.

Existing research works in resource utilization and energy management are good for ordinary cloud environment, but it does not suit for Hadoop-based cloud requests. The proposed work aims to provide an efficient scheduler that reduces energy consumed and distance between nodes. This can enhance performance of Hadoop in cloud and reduces energy consumption for cloud service providers. The factors considered for scheduling decisions are volume, compute capacity, CPU utilization of nodes, and the distance between data nodes in cluster.

3 Scheduling in OpenStack

Scheduling in an OpenStack environment involves placing VMs on suitable physical machines. When Hadoop cluster is to be set up in a cloud environment, the following issues are to be considered:

(a) On which physical node should the VM be launched?
(b) Which host must provide the volume required by the launched instance?

The first one is called a compute request and the second is called a volume request. In OpenStack, it is the responsibility of the nova scheduler to service volume and compute request. Chance scheduler [10] services volume requests by selecting host randomly that is active at the moment. Filter scheduler [10] services compute requests. When the filter scheduler receives a request, it first applies filters to determine which hosts are eligible for consideration. Hosts that are accepted by the filter are then weighted to decide which hosts to use for that request.

In OpenStack, no methods are implemented for placing nodes in energy-efficient manner. Also distances between nodes are not considered for scheduling. These factors are important to set up an energy-efficient Hadoop cluster in cloud environment.

4 Proposed Approach

The proposed work involves designing a new energy-efficient scheduler for the OpenStack cloud that will be efficient for Hadoop clusters. When the request for an instance to form a Hadoop cluster enters, the scheduler present in the cloud does the following:

- It chooses the nodes that have volume capability in addition to compute capability.
- The instances required must be scheduled in such a way that they are at a closer proximity to each other.
- It checks the energy consumed when VMs are launched in different nodes.

4.1 Energy-Efficient Scheduler

Optimal VM placement is an effective means to manage resources particularly in cloud environment. The scheduler uses the fact that processor utilization has a linear relationship with energy consumption. Processor utilization can be calculated with the number of instances or number of vCPUs assigned to a server. Utilization Ui is given in Eq. (1):

$$U = (\text{active vms})/(\text{Total number of vms}) \tag{1}$$

where active vms refers to the number of VMs running at the given time and total number of vms refers to the total number of VMs possible in that resource. Power supply efficiency is given in Eq. (2):

$$\text{Efficiency (n)} = (\text{P out})/(\text{P in}) \tag{2}$$

Energy-efficient scheduler makes decisions based on resource utilization, which is a key indicator for energy efficiency. This function aims to increase consolidation density, and its advantage is energy consumption is reduced. And the second advantage is that energy-efficient scheduler implicitly decreases the number of active resources since it tends to intensify the utilization of a small number of resources.

4.2 Distance Matrix

A distance matrix is maintained which contains the distance between the nodes. The distance between the two nodes is calculated by summing the distances to their closest ancestor. The distance between the same nodes is considered as 0 (distance (/d1/r1/n1, /d1/r1/n1) = 0). The distance between different nodes in the same rack is taken as 2 (distance (d1/r1/n1, /d1/r1/n2) = 2). The distance between different nodes in different racks but within the same data center is considered to be 4 (distance (/d1/r1/n1, /d1/r2/n3) = 4). The distance between different nodes in different racks and in different data centers is taken to be 6 (distance (/d1/r1/n1, /d2/r3/n4) = 6).

When a request comes in, the group of nodes that have minimum distance between each other is chosen. This is because shorter distance leads to greater bandwidth.

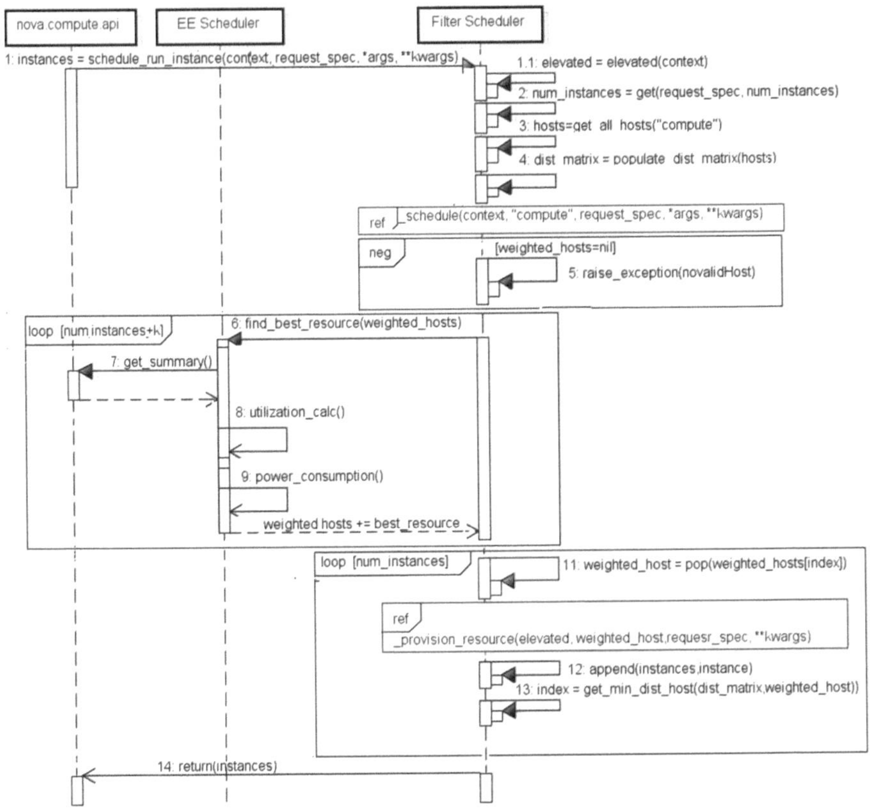

Fig. 2 Working of energy-efficient scheduler

5 Experimental Analysis

The cloud was set up in servers in which one has all OpenStack components including nova–compute and the other has only the nova–compute component [11]. Energy Star is given a Power and Performance Data Sheet [12] for Dell PowerEdge R210 (250W Power Supply) which is considered by our scheduler.

The filter scheduler was modified as shown in Fig. 2 to incorporate the proposed methodology. Before calling the _schedule function, the list of hosts in the compute queue is obtained and a distance matrix is formed that contains the number of hops between the hosts available. Before appending the current host to the list of selected hosts, the number of disks available in that host is acquired from host_state. capabilities. It is verified whether the host has sufficient amount of disk space.

After the sorted list is obtained from _schedule function, the list is given to the energy-efficient scheduler where the utilization of each physical machine and power consumption is calculated. The scheduler chooses the best resource based on calculated utilization rate and minimum power consumption required by the host

Table 1 Energy consumption by filter and EE scheduler

Scheduler	Energy consumed (server-I)	Energy consumed (server-II)	Total energy consumption
Filter scheduler	197.80 W	197.80 W	395.60 W
EE scheduler	197.34 W	69.1 W	266.44 W
Total energy saved is 129.16 W			

to run that instance. The number of best resources found out is greater than the requested number of resources. From the list of best resources, invariably the first host is popped off and returned. For the remaining number of instances, the host that has the minimum distance to the host that was previously allocated is chosen. This is donc with the help of the distance matrix framed earlier.

6 Result and Discussion

The working of filter scheduler is checked by launching instances in OpenStack. The configuration of filter scheduler is modified to implement energy-efficient scheduler.

Using filter scheduler, VMs are launched on different servers (instance runs on both server-I and server-II). Using EE scheduler, instances are launched on the same server-I to reduce distance between nodes and to reduce power consumption. Therefore, the number of active resources is reduced; the unused resource is idle so energy consumption is reduced as shown in Table 1. We achieved maximum resource utilization, placing the VM on node that has volume attached to it.

7 Conclusion

The proposed work aims at enhancing the features of scheduler to set up an energy-efficient Hadoop cluster in a private or public cloud. It overcomes the disadvantage of separation of storage and compute while setting up Hadoop cluster in the same node of the cloud. Further the scheduler considers the proximity of the machines that form Hadoop cluster. Energy-efficient placement of virtual machines is also considered based on the CPU utilization. Hence the unused physical machines can be switched off or moved to sleep state. A minimum of 36 % saving in energy is achieved using the proposed approach. The proposed work can be extended considering SLA and choosing suitable evolutionary techniques for optimization in large-scale cloud environment.

Acknowledgments We would like to express our sincere gratitude to Dr. R. Rudramoorthy, Principal, PSG College of Technology, for providing us with the necessary facilities for the work. We would also like to thank Dr. R. Venkatesan, HOD, Dept. of Computer Science and Engineering, for all the support provided to accomplish the work. The authors thank Mr. Chidambaran Kollengode, Director, Nokia R&D. This project is a PSG-Nokia Collaborative research work.

References

1. Cloud computing. en.wikipedia.org/wiki/Cloud_computing (25 Aug 2013)
2. Introduction to OpenStack and its components. http://docs.openstack.org/ (20 Sept 2013)
3. Tom, W.: Hadoop: The Definitive Guide. O'Reilly Media, Beijing (2012)
4. Pei Fan, Ji Wang, Zibin Zheng, Lyu, M.R.: Toward optimal deployment of communication-intensive cloud applications. In: IEEE International Conference on Cloud Computing, pp. 460–467. (2011)
5. Gihun Jung, Kwang Mong Sim: Location-aware dynamic resource allocation model for Cloud computing environment. In: International Conference on Information and Computer Applications, Hong Kong, pp. 37–41. (2012)
6. Beloglazov, A., Buyya, R., Lee, Y.C., Zomaya, A.: A taxonomy and survey of energy-efficient data centers and Cloud computing systems. Technical Report, Cloud Computing and Distributed Systems Laboratory, (2010)
7. Ching-Hsien Hsu, Slagter, K.D., Shih-Chang Chen, Yeh-Ching Chung: Optimizing energy consumption with task consolidation in clouds. Elsevier: Inf. Sci. **258**, 452–462 (2014)
8. Beloglazov, A., Abawajy, J., Buyya, R.: Energy-aware resource allocation heuristics for efficient management of data centers for Cloud computing. Elsevier: Future Gener. Comput. Syst. **28**, 755–768 (2012)
9. Gayathiri, N., Sudha Sadasivam, G.: Evaluation of hadoop on local clusters and Amazon public cloud. In: International Conference on Cloud and Big Data Analytics, India (2013)
10. Yong Sheng Gong: OpenStack nova-scheduler and its algorithm. In: IBM Developers work (2012)
11. Pepple, K.: Deploying openstack. Oreilly, (2011)
12. Power and Performance data sheet. http://www.dell.com/downloads/global/products/pedge/en/PowerEdge_R210_250W_Energy_Star_DataSheet.pdf (20 Nov 2013)

FTMXT: Fault-Tolerant Immediate Mode Heuristics in Computational Grid

Sanjaya Kumar Panda, Pabitra Mohan Khilar, and Durga Prasad Mohapatra

Abstract Fault tolerance plays a key role in computational grid. It enables a system to work smoothly in the presence of one or more failure components. The components are failing due to some unavoidable reasons like power failure, network failure, system failure, etc. In this chapter, we address the problem of machine failure in computational grid. The proposed system model uses the round trip time to detect the failure, and it uses the checkpointing strategy to recover from the failure. This model is applied to the traditional immediate mode heuristics such as minimum execution time (MET) and minimum completion time (MCT) (defined as MXT). The proposed Fault-Tolerant MET (FTMET) and Fault-Tolerant MCT (FTMCT) heuristics (defined as FTMXT) are simulated using MATLAB. The experimental results are discussed and compared with the traditional heuristics. The results show that the proposed approaches bypass the permanent failure and reduce the makespan.

Keywords Immediate mode • Minimum execution time • Minimum completion time • Scheduling • Fault tolerance • Grid computing

1 Background

Computational grid is widely used for high-performance computing applications [1]. It solves the large-scale complex problems by sharing and aggregation of resources [2]. The resources in grid are distributed worldwide and it is under different domains. One domain may have one or more instances. In order to

S.K. Panda (✉)
Department of CS&E, ISM Dhanbad, Dhanbad, Jharkhand, India
e-mail: sanjayauce@gmail.com

P.M. Khilar • D.P. Mohapatra
Department of CS&E, NIT Rourkela, Rourkela, OD, India
e-mail: pmkhilar@nitrkl.ac.in; durga@nitrkl.ac.in

E.B. Rajsingh et al. (eds.), *Informatics and Communication Technologies for Societal Development*, DOI 10.1007/978-81-322-1916-3_11, © Springer India 2015

maintain this, the Grid Referral Service (GRS) keeps track of all information about the domain and its instances [3]. It provides information to the Grid Machine/ Resource Broker (GMB/GRB). However, the broker is responsible for mapping the jobs and the available resources. It also splits the job into a number of small units called task. The scheduling model is shown in Fig. 1. The grid consumer submits the job to the broker. Then, the broker obtains the available machine list information from the GRS [4]. Finally, it maps the jobs to the resources based on the user requirements (i.e., task machine lists) and gets back the results (responses). The timeline sequence is shown in Fig. 2. In the grid, the heuristics are categorized into two types: immediate and batch mode. Immediate mode heuristic assigns a task to a resource as soon as the task arrives. Batch mode heuristic assigns a group of tasks to the resources at prescheduled times [5]. As resources are under different domains (refer Fig. 1), it may enter or leave at any point of time. Sometimes, it leads to burden on the grid [6]. Furthermore, the unpredictable resources are changing over time [7]. It may be interrupted by the domain administrator or fail due to some unavoidable circumstances. It drastically impacts on the scheduling as well as makespan.

Moreover, there are three types of failure in task scheduling. They are transient failure, intermittent failure, and permanent failure [8]. In this chapter, we have proposed a heuristic to handle the permanent failure and fail-stop model [9].

The rest of this chapter is organized as follows: Section 2 briefly discusses related research in fault tolerance and grid scheduling. Section 3 gives the preliminaries. In Sect. 4, two proposed heuristics are discussed and pseudocodes are presented. We conduct experiments and discuss results in Sect. 5. We conclude in Sect. 6.

2 Related Work

Many researchers have proposed various heuristics to reduce the makespan and increase resource utilization. But few authors address the problem of fault tolerance [10–14]. The problem of fault tolerance is of two types: resource fault (or failure) and network fault (or failure). In resource failure, the resource is not able to execute any task. But in network failure, the task is not able to arrive at the resource. Until the broker gets back the result, it may not possible to predict the failure, i.e., machine or network failure [15].

Nazir et al. [10] present the problem of fault tolerance in the form of machine failure. In this scheme, the grid information service maintains the history of fault occurrence. The value of the resource fault occurrence index is ranging from 0 to 6. If the fault index is zero, then the job is assigned to the resource. If it is in between 1 and 3, then checkpoint is appended at 50 %. So it returns the result to the broker after the 50 % execution. Similarly, if it is in between 4 and 6, then checkpoint is appended at 20 % because there is a high chance of failure.

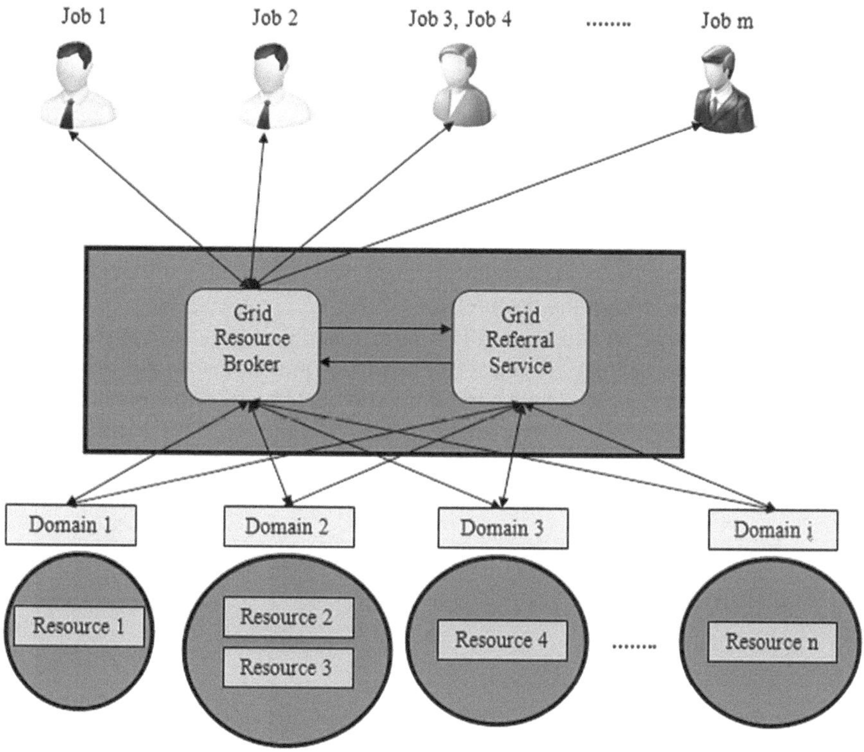

Fig. 1 A scheduling model

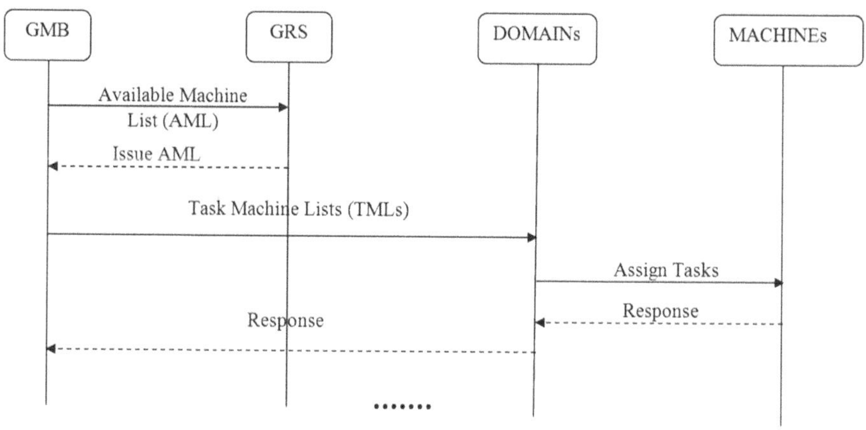

Fig. 2 Timeline sequence

Khanli et al. [12] maintain a fault occurrence history in grid information service. It uses genetic algorithm to schedule the jobs and checkpointing strategy to make scheduling more efficient. Upadhya et al. [13] propose a fault-tolerant technique based on checkpointing and passive replication. It uses genetic algorithm to perform the scheduling. To achieve reliability in a grid system, Guo et al. [11] introduce local node fault recovery mechanism. This mechanism uses ant colony optimization to solve multi-objective task scheduling. A two-phase load balancing algorithm is introduced by Nanthiya et al. [14]. First, resource selection is performed using deadline and its fault tolerance factor. Second, the load balancing algorithm is used to assign the resources.

In this chapter, we have proposed two immediate mode heuristics using round trip time and checkpointing concept. The heuristics are applied to the benchmark data set by Braun et al. [16]. To the best of our knowledge, this is the first immediate mode fault tolerance work using the benchmark data set. This chapter is an extension of our previous work [17, 18].

3 Preliminaries

3.1 Problem Statement

Assume that a grid scheduler has m different types of jobs (or tasks) $J_1, J_2, \ldots, J_m$ and n different types of resources (or machines) $R_1, R_2, \ldots, R_n$. The problem is to design an efficient heuristic to map all jobs to the available resources such that the overall processing time is minimized. In addition to this, if a resource has failed due to some unavoidable circumstances, then the proposed heuristic is able to handle it. In this chapter, we have considered the problem as the fault tolerance problem in computational grid.

3.2 Scheduling Algorithms

There are many heuristics in immediate mode heuristic. But we have listed only two heuristics: MET and MCT.

3.2.1 MET

It maps a task to a resource which takes least execution time. The main demerit of this heuristic is load imbalance because it is not considering the resource load. If the least execution time resources have failed due to some unavoidable circumstances, then this heuristic will not execute a single task. This heuristic takes O(n) time to map a task to a resource [5].

3.2.2 MCT

It maps a task to a resource which takes least completion time. The completion time is the sum of the execution time and resource ready time. This heuristic is a benchmark for immediate mode heuristic. But this heuristic is not considering the fault tolerance aspects. This heuristic takes O(n) time to map a task to a resource [5].

4 Proposed Heuristics

4.1 Description

The proposed heuristics are divided into two phases: matching and scheduling. The matching phase is similar to the traditional MET (or MCT) heuristic. The proposed fault-tolerant technique is introduced in scheduling phase.

4.2 Heuristics

The main mechanism of FTMET heuristic is defined in Algorithm 1. This heuristic maps a task to a machine which takes least execution time (lines 1 to 3). Thereafter, the broker finds the current status of the machine from GRS. Then, the GRS checks whether the machine is able to compute a task or not. It sends a message to the machine and waits for a confirmation message. If GRS will not get back the confirmation message in stipulated time, it concludes that the machine has failed or is unable to respond to the request. For that, it calculates round trip time (RTT), which is the sum of the time to send a message (to the machine) and acknowledge for it. Finally, the GRS informs the broker about the current status of the machine, and the GMB reschedules the task to the next least execution time machine. Then again, it finds the current status of the machine from GRS. If the machine works normally, then the task is assigned to the machine. Lines 4 to 12 show the scheduling process. The above process is also followed in our previous work [17, 18]. In addition, the mathematical representation is also shown over there.

Checkpointing strategy is used to recover the failure. If timeout occurs, then the GRS requests the broker to roll back to the last consistent state. Furthermore, the GRS is not sending any task to that failed (or timeout) machine.

Algorithm 1. FTMET heuristic

1. **for** task T_i
2.   **for** all machine M_j
3.   Find minimum $E_{i,j}$ and machine M_j that holds it. // $E_{i,j}$ = Execution Time
4.   Set K = 1.
5.   Find the status of machine M_j from GRS.
6.   **if** (M_j == Faulty)
7.   Find (K + 1) minimum $E_{i,j}$ for task T_i and machine M_j that holds it.
8.   Go to Step 5.
9.   **else** Assign task T_i to machine M_j.
10.   **end if**
11.   **end for**
12. **end for**

The main mechanism of FTMCT heuristic is defined in Algorithm 2. This heuristic maps a task to a machine which takes least completion time (lines 1 to 7). Thereafter, the broker finds the current status of the machine from GRS. If the machine fails, then the broker reschedules the task to the next least completion time machine, and it again finds the current status of the machine from GRS. Otherwise, the task is assigned to the machine. Lines 8 to 15 show the scheduling process. Like FTMET, this heuristic uses RTT and the checkpointing method in similar fashion.

Algorithm 2. FTMCT heuristic

1. **for** task T_i
2.   **for** all machine M_j
3. $C_{i,j} = E_{i,j} + R_j$ // $C_{i,j}$ = Completion Time, R_j = Ready Time
4.   **end for**
5. **end for**
6. **for** task T_i
7.   Find minimum $C_{i,j}$ and machine M_j that holds it.
8.   Set K = 1.
9.   Find the status of machine M_j from GRS.
10.   **if** (M_j == Faulty)
11.   Find (K + 1) minimum $C_{i,j}$ for task T_i and machine M_j that holds it.
12.   Go to Step 9.
13.   **else** Assign task T_i to machine M_j.
14.   **end if**
15. **end for**

Table 1 Expected execution time of four tasks on three machines

Task	M_1	M_2	M_3
T_1	120	75	32
T_2	40	110	93
T_3	71	24	49
T_4	23	34	47

5 Experimental Study

In this section, we present the results in terms of two performance measures makespan and machine utilization. Makespan is the overall completion time taken to assign all tasks to the machines. Machine utilization is the percentage of time the machine is busy. The results are carried out using MATLAB.

Let us consider a 4×3 matrix shown in Table 1. We assumed that tasks arrive in the following order: T_1, T_2, T_3, and T_4. In MET, task T_1 has least execution time on machine M_3. So it is assigned to machine M_3. Like task T_1, the least execution time for task T_2, task T_3, and task T_4 is machine M_1, machine M_2, and machine M_1, respectively. So the tasks are assigned to the machine accordingly. The overall makespan is 63.

Consider a scenario in which the machine M_1 has failed due to some unavoidable circumstances. So task T_2 and task T_4 are not executed successfully. In the proposed FTMET, if the machine M_1 has failed due to some unavoidable circumstances, then task T_2 and task T_4 are assigned to the second least execution time machine, i.e., machine M_3 and machine M_2, respectively. The overall makespan is 125. The major advantage is to keep the task against faults.

In MCT, task T_1 has least completion time on machine M_3. So it is assigned to machine M_3. However, for task T_2, the least completion time machine is M_1. So task T_2 is assigned to machine M_1. Similarly, task T_3 and task T_4 are assigned to machine M_2. The overall makespan is 58. Note that MET/MCT scheduling assumes that none of the machines have failed in the middle of the execution.

Consider a scenario in which the machine M_1 has failed due to some unavoidable circumstances. So the task is not executed successfully. In the proposed FTMCT, if the machine M_1 has failed due to some unavoidable circumstances, then task T_2 is assigned to the second least completion time machine, i.e., machine M_2. The overall makespan is 128.

We have considered Braun et al. [16] data sets (or instances) to evaluate the proposed heuristics. The instances are classified into 12 different types of matrices. The general form of the instance is u_t_mmnn.o. Here, u indicates the uniform distribution; t indicates the nature of the matrix: consistent (c), inconsistent (i), and semi-consistent (s); mm indicates the task heterogeneity; and nn indicates the machine heterogeneity. The value of mm or nn is either hi or lo.

In this chapter, we have taken three different sizes of the data sets: 512×16, 1024×32, and 2048×64. In 512×16 instances, we have assumed that eight numbers of machines have failed during simulation. The machine IDs are M_{10}, M_3, M_4, M_{15}, M_1, M_8, M_{16}, and M_6. The machines are failed after the execution of the task ID T_{165}, T_{176}, T_{182}, T_{188}, T_{234}, T_{314}, T_{338}, and T_{370}, respectively.

Table 2 Numerical results of makespan value for 512 × 16, 1024 × 32, and 2048 × 64 instances, respectively

Instance	MET (512×16)	MET (1024×32)	MET (2048×64)	FTMET (512×16)	FTMET (1024×32)	FTMET (2048×64)	MCT (512×16)	MCT (1024×32)	MCT (2048×64)	FTMCT (512×16)	FTMCT (1024×32)	FTMCT (2048×64)
u_c_hihi	2.2159E+07	3.8431E+07	1.6736E+07	5.3052E+07	7.1204E+07	3.5770E+07	1.1423E+07	3.1749E+07	2.7362E+07	2.1794E+07	9.0368E+07	7.2217E+07
u_c_hilo	5.3951E+05	3.6308E+06	1.6641E+06	8.7844E+05	7.3878E+06	3.5821E+06	1.8589E+05	3.1614E+06	2.6695E+06	3.1088E+05	9.3558E+06	7.1534E+06
u_c_lohi	6.6846E+05	3.2742E+03	1.7626E+03	1.6256E+06	6.8098E+03	3.7223E+03	3.7830E+05	2.8765E+03	2.7639E+03	7.3675E+05	9.1600E+03	7.3549E+03
u_c_lolo	1.8065E+04	3.7785E+02	1.7265E+02	2.8962E+04	7.8480E+02	3.7123E+02	6.3601E+03	3.2576E+02	2.7196E+02	1.0175E+04	9.2257E+02	7.2791E+02
u_i_hihi	3.7073E+06	6.7612E+06	4.1277E+06	8.5274E+06	1.3968E+07	5.2893E+06	4.4136E+06	7.4194E+06	3.6175E+06	9.2631E+06	1.5616E+07	5.6261E+06
u_i_hilo	9.4796E+04	6.7070E+05	4.6574E+05	1.6403E+05	1.4525E+06	5.7573E+05	9.4856E+04	6.7008E+05	4.0982E+05	1.6698E+05	1.5199E+06	5.7434E+05
u_i_lohi	1.4232E+05	8.5439E+02	4.2063E+02	2.7597E+05	1.6435E+03	5.2275E+02	1.4382E+05	7.5134E+02	3.8518E+02	3.0498E+05	1.5039E+03	5.6427E+02
u_i_lolo	5.3993E+03	9.1120E+01	3.4820E+01	5.6228E+03	1.4735E+02	6.1960E+01	3.1374E+03	6.9460E+01	4.0810E+01	5.2437E+03	1.5151E+02	5.7580E+01
u_s_hihi	1.1077E+07	2.4737E+07	9.8003E+06	3.0105E+07	4.0502E+07	2.2545E+07	6.4227E+06	1.7347E+07	1.5599E+07	1.5219E+07	5.5970E+07	4.0392E+07
u_s_hilo	2.7135E+05	2.2116E+06	8.2527E+05	3.1881E+05	3.7105E+06	1.7650E+06	1.1837E+05	1.7473E+06	1.3726E+06	2.3047E+05	5.4920E+06	3.7840E+06
u_s_lohi	3.0255E+05	2.1260E+03	8.7390E+02	8.6459E+05	3.6559E+03	1.7365E+03	1.8409E+05	1.6444E+03	1.3767E+03	4.4428E+05	5.4971E+03	3.9021E+03
u_s_lolo	3.6922E+03	1.7873E+02	8.5660E+01	1.2084E+04	4.0626E+02	2.1785E+02	4.4361E+03	1.8050E+02	1.4440E+02	8.1171E+03	5.7442E+02	4.0979E+02

Table 3 Numerical results of machine utilization value for 512 × 16, 1024 × 32, and 2048 × 64 instances, respectively

Instance	MET (512×16)	MET (1024×32)	MET (2048×64)	FTMET (512×16)	FTMET (1024×32)	FTMET (2048×64)	MCT (512×16)	MCT (1024×32)	MCT (2048×64)	FTMCT (512×16)	FTMCT (1024×32)	FTMCT (2048×64)
u_c_hihi	1	1	1	0.7088	0.6000	0.5709	0.7020	0.6475	0.6167	0.6424	0.5794	0.5957
u_c_hilo	1	1	1	0.8071	0.5580	0.5523	0.7090	0.6495	0.6168	0.6489	0.5668	0.5871
u_c_lohi	1	1	1	0.7056	0.5729	0.5494	0.7054	0.6480	0.6111	0.6320	0.5714	0.5826
u_c_lolo	1	1	1	0.8119	0.5558	0.5600	0.6980	0.6448	0.6196	0.6568	0.5758	0.5799
u_i_hihi	0.5741	0.4994	0.3781	0.4901	0.4437	0.4046	0.7077	0.6176	0.5816	0.6285	0.5609	0.5313
u_i_hilo	0.5711	0.4495	0.3522	0.5080	0.4117	0.3901	0.7016	0.6428	0.5679	0.6339	0.5570	0.5662
u_i_lohi	0.5002	0.3730	0.3855	0.5091	0.3956	0.4418	0.6961	0.6058	0.5897	0.6256	0.5855	0.5793
u_i_lolo	0.5587	0.3546	0.4494	0.4949	0.4196	0.3627	0.7202	0.6419	0.5687	0.6471	0.5785	0.5607
u_s_hihi	0.2420	0.0928	0.0953	0.3032	0.1337	0.0801	0.7119	0.6704	0.5808	0.6210	0.5671	0.5746
u_s_hilo	0.2660	0.1116	0.1208	0.4960	0.1447	0.0935	0.7339	0.6471	0.6029	0.6264	0.5591	0.5848
u_s_lohi	0.2863	0.1137	0.1094	0.3105	0.1361	0.0918	0.7103	0.6696	0.6145	0.6060	0.5439	0.5607
u_s_lolo	0.3109	0.1184	0.1083	0.4637	0.1226	0.0809	0.7093	0.6317	0.6087	0.6227	0.5527	0.5610

The numerical results of makespan value for 512×16, 1024×32, and 2048×64 instances are shown in Table 2. Similarly, the numerical results of machine utilization value for the above instances are shown in Table 3 respectively. The RTT (or timeout) and checkpointing time overhead are negligible for simplicity. In Tables 2 and 3, MET and MCT heuristics are very less makespan because very few number of tasks are executed (due to machine failure). For example, let us consider the 512×16 instances. In MET, the total number of tasks executed in u_c_hihi instances is 512–279 (total number of failed task) = 233. The makespan is 2.2159E + 07. But the proposed FTMET executes 512 tasks in the presence of failure. Hence, the makespan is 5.3052E + 07. So FTMET gives better performance in the presence of failure.

6 Conclusion

In this chapter, we have introduced two new heuristics: FTMET and FTMCT. These heuristics are an extension of the existing MET and MCT heuristic. The proposed heuristics address the problem of fault tolerance and its solution. Heuristics are evaluated using benchmark instances. The results show that the proposed heuristics give better task scheduling and minimize the makespan.

References

1. Medeiros, R., Cirne, W., Brasileiro, F., Sauve, J.: Faults in Grids: why are they so bad and what can be done about it. In: Proceedings of the Fourth International Workshop on Grid Computing. (2003)
2. Murshed, M., Buyya, R., Abramson, D.: GridSim: A Toolkit for the Modeling and Simulation of Global Grids, pp. 1–15. Monash University Journal. (2001)
3. Vasques, J., Veiga, L.: A decentralized utility-based grid scheduling algorithm. In: 28th Annual ACM Symposium on Applied Computing, pp. 619–624. (2013)
4. Li, M., Xiong, N., Yang, B., Li, Z., Park, J.H., Lee, C.: Posted price model based on GRS and its optimization for improving grid resource sharing efficiency. Telecommun. Syst. **55**(1), 71–79 (2014)
5. Maheswaran, M., Ali, S., Siegel, H.J., Hensgen, D., Freund, R.F.: Dynamic mapping of a class of independent tasks onto heterogeneous computing systems. J. Parallel Distrib. Comput. **59**, 107–131 (1999)
6. Sadashiv, N., Kumar, S.M.D.: Cluster, grid and Cloud computing: a detailed comparison. In: IEEE 6th International Conference on Computer Science and Education, Singapore, pp. 477–482. (2011)
7. Ergu, D., Kou, G., Peng, Y., Shi, Y., Shi, Y.: The analytic hierarchy process: task scheduling and resource allocation in cloud computing environment. J. Supercomput. **64**, 835–848 (2013). Springer
8. Mushtaq, H., Al-Ars, Z., Bertels, K.: Survey of fault tolerance techniques for shared memory multicore/multiprocessor systems. In: IEEE 6th International Design and Test Workshop, Beirut, Lebanon, pp. 12–17. (2011)

9. Treaster, M.: A Survey of Fault-Tolerance and Fault-Recovery Techniques in Parallel Systems. National Center for Supercomputing Applications. University of Illinois. (2005)
10. Nazir, B., Khan, T.: Fault tolerant job scheduling in computational grid. In: IEEE 2nd International Conference on Emerging Technologies, Peshawar, Pakistan, pp. 708–713. (2006)
11. Guo, S., Huang, H., Wang, Z., Xie, M.: Grid service reliability modeling and optimal task scheduling considering fault recovery. IEEE Trans. Reliab. **60**, 263–274 (2011)
12. Khanli, L.M., Far, M.E., Rahmani, A.M.: RFOH: a new fault tolerant job scheduler in grid computing. In: IEEE 2nd International Conference on Computer Engineering and Applications, Bali, Indonesia, pp. 422–425. (2010)
13. Upadhyay, N., Misra, M.: Incorporating fault tolerance in GA-based scheduling in grid environment. In: IEEE World Congress Information and Communication Technologies, Mumbai, India, pp. 772–777. (2011)
14. Nanthiya, D., Keerthika, P.: Load balancing GridSim architecture with fault tolerance. In: International Conference on Information Communication and Embedded Systems, Chennai, India, pp. 425–428. (2013)
15. Duarte, E.P., Weber, A., Fonseca, K.V.O.: Distributed diagnosis of dynamic events in partitionable arbitrary topology networks. IEEE Trans. Parallel Distrib. Syst. **23**, 1415–1426 (2012)
16. Braun, T.D., Siegel, H.J., Beck, N., Boloni, L.L., Maheswaran, M., Reuther, A.I., Robertson, J. P., Theys, M.D., Yao, B.: A comparison of eleven static heuristics for mapping a class of independent tasks onto heterogeneous distributed computing systems. J. Parallel Distrib. Comput. **61**, 810–837 (2001)
17. Panda, S.K., Khilar, P.M., Mohapatra, D.P.: FTM^2: fault tolerant batch mode heuristics in computational grid. In: 10th International Conference on Distributed Computing and Internet Technology. Lecture Notes in Computer Science, vol. 8337, pp. 98–104. (2013)
18. Panda, S.K.: Efficient scheduling heuristics for independent tasks in computational grids. M. Tech. thesis, National Institute of Technology Rourkela (2013)

Assurance System in Traditional Body Area Network for Medical Diligence

Ankit Agarwal, Burla Rajesh, and K. Aanandha Saravanan

Abstract Due to the rapid progression of wireless engineering systems, a special domain of sensor network which supports medical diligences, is the development of the autonomous wireless body area network. In this type of environment it is useful for things such as monitoring the health condition of senior citizens. The sensed signal can monitor several types of measurements such as electroencephalography (EEG), electrocardiography (ECG), electromyography (EMG), skin temperature, skin conductance or electrooculography (EOG) etc., it then compares with the standard level and the information is passed to the care takers or hospital for further necessary actions using a hybrid network of cryptography which nowadays is crucial. Privacy information is vital and it must be safe guarded from several aggressors. We propose a method of crypt analyst method using a cluster pattern of key mechanism which provides an assured, efficient and improved way to increase the lifetime of the network.

Keywords Wireless body area network (WBAN) • Cryptography • Key mechanism • Assure • Efficient

1 Introduction

Wireless sensor networks (WSN) are in increasing demand and can be a tiny sensing material that is self-powered which gathers information or detects special events and communicates in a wireless pattern with the end goal of handing the processed data to a base station. Its senses the information, processes it and communicates through any medium. These tiny devices are capable of detecting various application fields such as automation, logistics and transportation, environmental monitoring,

A. Agarwal (✉) • B. Rajesh • K.A. Saravanan
Department of ECE, Veltech University, Chennai, India
e-mail: ankit060691@gmail.com; raaz.burla@gmail.com; eceanandkumar@gmail.com

E.B. Rajsingh et al. (eds.), *Informatics and Communication Technologies for Societal Development*, DOI 10.1007/978-81-322-1916-3_12, © Springer India 2015

healthcare, security and surveillance. Sensor devices for monitoring human activities have generated great research interest in recent years [1]. Their application areas include two major subdivisions—non-medical practice and medical practice. A non-medical device and its application could be real time video streaming using video players, real time audio streaming using audio players remote control application, data file transfer, sports and gaming. Medical applications include wearable and implantable devices [2]. Wearable devices can be used for close proximity of the surface of a human body; they include a respiration monitor, temperature measurement, heart rate monitor, pulse oximeter, blood pressure monitor, pH monitor, glucose sensor etc. Implantable devices are inserted in the human bodies and their application includes brain liquid pressure sensor, endoscope capsule, cardiac arrhythmia monitor/recorder etc. Apart from these monitoring applications there are some applications for measuring body positions and their location, overall monitoring of ill patients and hospitals and homes.

The monitoring of human organs has given rise to a new field of sensor technology known as wireless body area networks (WBAN). A WBAN is a radio frequency based wireless networking technology that interconnects tiny sensor or actuator capabilities in, on or around a human body covering transmission of these nodes over a short range of about 2 m. The applications of WBANs include monitoring of human health remotely by implanting the biometric sensor on the human body. WBAN needs few sensor nodes to collect physiological data from the body since it has a single centralized entity called identity server (IS), that collects data from the sensor nodes using multi-hop communication and sends it to a remote medical server which is connected to the care taker or hospital. The IS acts as a gateway between the body on which it resides and the outside world. These devices provide continuous health monitoring and real-time feedback to the user or medical personnel. Moreover, these measurements can be recorded over a longer period of time [3]. WBANs communications are divided into two parts; intra-WBAN communication and inter-WBAN communication. The on body communication between sensor nodes comes under intra-WBAN communication whereas communication occurring between IS residing on two or more bodies comes under inter-WBAN communication. Similarly, inter-WBAN communication can be taken for the health monitoring system, for example monitoring the patients' health while they are involved in their normal routine work (likewise somewhere in market, home, office or even in the playground). In this case WBANs (sensor clothed bodies) are scattered and the sink base station (SBS) might not always be in their range. So WBANs must cooperate among them using hop-by-hop communication to reach SBS, and SBS communicates further to remote central medical server (RCMS) through the internet. Figure 1 below gives an idea about basic WBAN through remote services.

The architecture of WBAN is divided into three levels.

- The lowest level consists of a set of intelligent sensors or nodes which are the reduced function devise. In the non-medical field, wearable devices such as a headset, an mp3 player, a game controller, etc., can be included in this level.

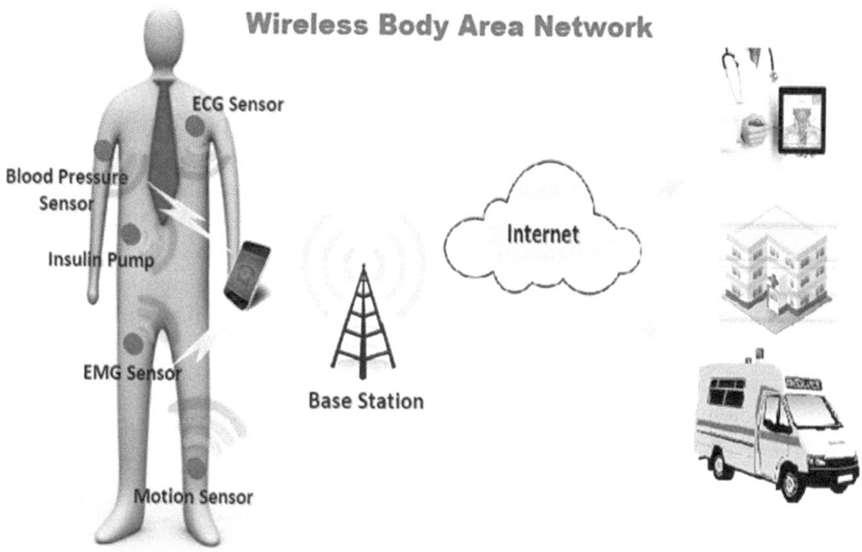

Fig. 1 Architecture of WBAN

- The second level is the mobile personal server that has full function devices like internet enabled PDA or a cell-phone. It communicates with the external network's lowest level devices. Additionally, it can display the analysed information on a screen with an option.
- The third level includes an external network of remote servers which provides various application services. For example, the medical server keeps electronic medical records of registered users and provides various services to the users, medical personnel and informal caregivers.

2 Related Works

Research work in health care applications of sensor devices in the security of the wireless communication are under progress over many parts of the world. Numerous projects are developed and are undergoing different stages such as wearable health devices. Funds are raised by both the government sector and private sector. Clinical prototypes for implantable and wearable health monitoring sensors have already been designed in recent years. These devices are capable of monitoring the human body for longer periods of time. Most of the work is focused on the concern related to biocompatibility, power-efficiency and reliability [4]. Therefore, securing communication between these devices is of great concern.

3 Proposed Model

The proposed scheme uses a non-probabilistic cluster based key management scheme for WBANs communication which is of higher security than a simple key management schemes. A cluster based key management scheme has a pool of sensors interconnected, one cluster head and the rest cluster members. The cluster head communicates with the cluster head of another pool. A simple WBAN may contain several pools of cluster that require inter communication among them proposing different cluster algorithms for each communication. Clustering approaches can be of two types—probabilistic and non-probabilistic. In this paper, we propose a method of key management to exchange their identity between the inter and intra network using hash function. This type of architecture provides an energy efficient reliable network. The model we propose uses a three tier architecture with elliptic curve digital signature algorithm and signature vapour and Diffie-Hellman algorithm with the effective background of clustering. This algorithm will be briefly explained in the next sections.

3.1 Clustering

Nowadays, in most wireless body area networks (WBAN) applications the entire network must have the ability to operate unattended in harsh environments in which pure human access and monitoring cannot be easily programmed or proficiently managed [5]. As a result, the specialized energy-aware routing and data gathering protocols which offer high scalability have to be applied in order that network lifetime is preserved acceptably high in such environments.

Naturally, grouping sensor nodes into clusters has been widely adopted by the research community to satisfy the above scalability objective and generally achieve high energy efficiency and prolong network lifetime in large-scale WBAN environments. The corresponding hierarchical routing and data gathering protocols imply cluster-based organization of the sensor nodes in order that data fusion and aggregation are possible, thus leading to significant energy savings [6, 7]. In the hierarchical network structure each cluster has a leader, which is known as the cell cluster head (CCH) which functions in special tasks such as fusion and aggregation, and several common beacon sensor nodes (BSN) as members (Fig. 2).

We use a non-probabilistic clustering algorithm for CCHs selection and cluster formation which are mainly based on the nodes. The cluster formation procedure depends on the communication of nodes with neighbours (one or multiple hops) and usually requires more intensive exchange of message and probably graph traversing in some amount. Furthermore, an even more challenging and promising non-probabilistic clustering approach is based on the use of cloud intelligence and has led to the construction of bio inspired clustering protocols to extend network lifetime. During the application of typical proximity-connectivity criteria, the

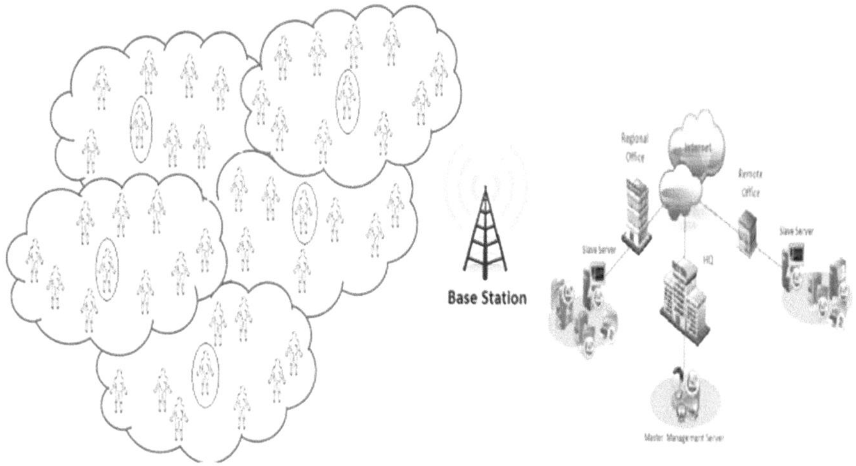

Fig. 2 Cluster model for WBAN

proposed protocols are primarily guided by specific structure-based sensor organization and then by progressive cluster formation steps which normally lead to a cluster with more control characteristics [8].

3.2 Key Management Using Clustering

Key pre-distribution in wireless sensor networks has attracted the attention of researchers for a decade. Key pre-distribution schemes are classified into two groups, viz. probabilistic key pre-distribution and deterministic key pre-distribution. In probabilistic key pre-distribution scheme, as the name implies, the keys are randomly drawn from a large pool of keys and are placed into the individual sensor nodes. This scheme does not ensure full connectivity between nodes. However, due to this scheme's randomness, it does ensure resiliency against selective node capture attack. The main disadvantage of probabilistic key pre-distribution schemes are that they do not ensure full connectivity between each and every pair of nodes [9]. On the other hand, in deterministic key pre-distribution scheme, a deterministic method is employed to load the keys into the sensor nodes. This scheme may or may not offer full connectivity between every pair of nodes of the wireless.

3.3 Three Tier Architecture

We have chosen Blundo scheme to construct our approach. Use of the Blundo scheme, therefore, greatly eases the presentation of our study and enables us to provide a clearer security analysis.

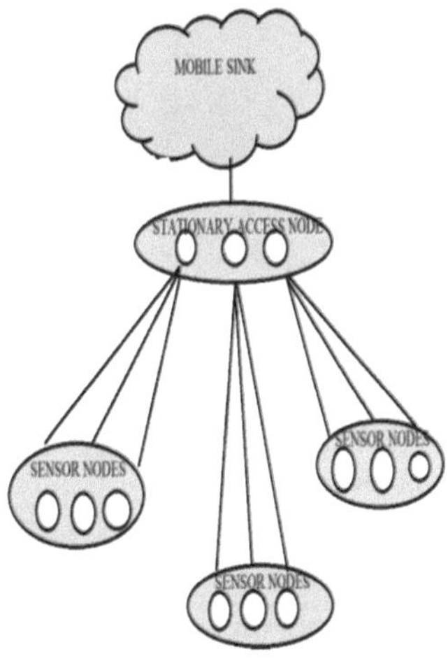

Fig. 3 Three tier architecture

In three tier architecture, the sensor nodes are grouped into two groups. The first group consist of normal *sensor* which performs the data acquisition or the data collection of the hostile environment. Secondly the sensor group of the data are *stationary access points or cluster heads* when this node collects data from the sensors which are grouped under it. The servers or the devices which rely on the data from the network are called *mobile sinks* (Fig. 3).

The three tier architecture is formed when the sensor group under one stationary access node is connected to the mobile sink, so when the mobile sink needs the data, it contacts the particular cluster heads which in turn collect the data from the sensors. Now if a sensor wants to transmit then it has two ways: (1) Firstly, direct key discovery, i.e. the sensor nodes contacts the access and exchanges the key with it which in turn contacts the mobile sink. (2) Secondly, indirect via stationary node [10], i.e. it contacts a stationary node which in turn contacts another stationary node which contacts the sink.

In this model we use two encryption methods, namely Diffie-Hellman algorithm and ECSDA to differentiate the two models. First we use an ECDSA algorithm, and signature vapour between mobile sink and stationary nodes; whereas the Diffie-Hellman algorithm and signature vapour is used in between the stationary node and sensors. This means that the data will be authenticated twice while it travels from sensor node to mobile sink. The algorithms will be discussed briefly in the following sections.

For clarity, we consider that one sender broadcasts messages to many receivers. In the network with multiple senders, each sender and its corresponding receivers are just the case we are considering. We consider that the entire sensor node has been given adequate power supply and has enough life cycle, we consider only one mobile sink, in a real life situation the number of mobile sinks is not limited [11]. Neither is the number of sensor nodes in each cluster nor the number of the clusters used. Our proposal can be used for any situation present in the real life world. The main advantage of the system is that an attacker has to know the private key to launch an attack which becomes difficult in the scenario. Now we move on to the next section explaining ECDSA.

3.4 Signature Vapour

The signature vapour is a technique to generate a self-sufficient message which when authenticated can lead to generation of the packet. This concept is used to reduce the generational computation time required to authenticate all the message, as the message will needed to be authenticated only once in the whole transmission of the whole packet. It can be described in three easy steps as follows.

3.4.1 Generating Extended Block

In this method, message M containing the '*n*' message is broken into '*k*' blocks of B_i, $1 < i < k$, then each block is broken into *b* messages m_j, $1 < j < b$. Now, each block of the message is concatenated using the function CON as shown below.

$$\mathrm{CON}(B_i) = m_{(i-1)\ b+1}\|\cdot\|m_{ib}, 1 \le i \le k \tag{1}$$

Then, CON (Bi), $1 \le i \le k$, is padded with authenticators, denoted by PAD (CON (Bi)) as shown in Eq. (2). CON (Bi), $1 \le i \le k-1$, is padded with digest $di+1$ which is the digest of PAD (CON ($Bj+1$)), as shown in Eq. (3) where H is a collision resistant hash. Note that CON (Bk) is padded with random characters which is denoted by $dk+1$ for consistency.

$$\mathrm{PAD}(\mathrm{CON}(Bi)) = \mathrm{CON}(Bi)\|di + 1 \tag{2}$$
$$dj + 1 = H(\mathrm{PAD}(\mathrm{CON}(Bj + 1))) \tag{3}$$

Block Bi and digest $di+1$ comprise an extended block EBi, i.e. $EBi = [Bi\ di + 1]$, $1 \le i \le k$. Let EBi, $1 \le i \le k$, comprise a vector $EB*$, i.e. $EB* = [EBi]Ti = 1,\ldots,k = [Bi\ di+1]Ti = 1,\ldots,k$. Because $EB*$ stands for an extended matrix of B, EBi is called an extended block. The authenticator is digest $d1$ that is the collision resistant hash value of PAD (CON ($B1$)), i.e. $d1 = H$ (PAD (CON ($B1$))). The authenticity of $d1$ is guaranteed by a signature generated by sender s with PRs. Thus, $EB0 = d1//E$ (PRs, $d1$). Until this point, all extended blocks have been generated. For clarity, let all extended blocks comprise a vector EB, i.e. $EB = [EBi]Ti = 0,\ldots,k$. Now when the signature in the first field is authenticated then the whole message is authenticated by the use of hash function.

3.4.2 Broadcasting the Extended Block

Authentication of each message depends on the arrival of the last packet which makes the authentication of the next message possible using the digest present in the first message [12]. Hence the messages have to be received in a sequential manner. We employ the sequential broadcast and reliable broadcast to guarantee the successful authentication of extended blocks with low overhead. They also ensure the integrity of *M* to applications.

3.4.3 Verification of the Extended Block

According to broadcasting extended blocks step, *EB*0 reaches receivers in *R* first. *d*1 in *EB*0 is authenticated by the signature, that is, if $D(PUs, E(PRs, d1)) = d1$, *d*1 is authentic. Extended blocks in $EB*$ are authenticated in an efficient way, just using a collision resistant hash. Digest *di*, $1 \leq i \leq k$, in $EBi - 1$ that reaches receivers in *R* in advance is used to authenticate *EBi*, that is, if $H(m(i - 1)b + 1 \parallel \ldots \parallel mib \parallel di + 1) = di$, *EBi* is authentic.

3.5 ECDSA

The ECDSA signature is on the basis of ECC, which offers equivalent security with substantially smaller key size compared to RSA. Thus, ECC has the advantages in computation, bandwidth and memory savings [13]. Because of the advantages, *d*1 in *EB*0 is signed with an ECDSA signature. Here, we introduce briefly generation and verification of the ECDSA signature.

3.6 Diffie-Hellman Algorithm

This algorithm uses a shared secret key concept on the data encryption and decryption, which is the main objective of the uses of the concept that $a*b = b*a$. Here the two users generate any two primes which are mutually exclusive.

3.6.1 Encryption

Now the data m is converted into integer using proper function then it is signed by sending as:

$$tx = mc^{pq}$$

3.6.2 Decryption

Now the data can be decrypted by:

$$m = txc^{-pq}$$

4 Security Analysis

We have analysed the performance of the proposed scheme using two metrics: security and connectivity [14]. For security, we present the probability of a mobile polynomial being compromised; hence, an attacker can make use of the captured mobile polynomial to launch a mobile sink replication attack against the sensor network. In connectivity, we estimate the probability Pconn:

$$\text{Pconn} = 1 - \left(1 - \frac{c}{n}\right)^m$$

n represents the total number of sensor nodes in the network, c is the average number of neighbour nodes for every sensor node before deployment of the stationary access nodes and m is the number of stationary access nodes in the network.

The probability that a mobile sink and a stationary access node share a information in a secured network is given by Pm

$$Pm = \frac{Km}{|M|}$$

The probability Ps, where two sensor nodes share a common static polynomial—the probability that the two sensors can establish a secure link directly is estimated by:

$$Ps = 1 - \frac{\left(\frac{|S|}{2Ks}\right) * \left(\frac{2Ks}{Ks}\right)}{\left(\frac{|S|}{Ks}\right) * \left(\frac{|S|}{Ks-1}\right)}$$

The probability Psa, where a sensor node and a stationary access node share a common static polynomial—the probability that the two nodes can establish a pairwise key directly is estimated by:

$$Psa = 1 - \frac{\left(\frac{|S|}{Ks-1}\right) * \left(\frac{Ks-1}{Ks}\right)}{\left(\frac{|S|}{Ks}\right)^2}$$

The probability Pa, where two stationary access nodes share a common static/mobile polynomial, is estimated by:

$$Pa = 1 - \frac{(|M| - 1)\left(\frac{|S|}{2(Ks-1)}\right) * \left(\frac{2.(Ks-1)}{Ks-1}\right)}{|M|.\left(\frac{|S|}{Ks}\right)^2}$$

The probability Pd of a mobile sink and a sensor node establishing a pair wise key (directly or indirectly) can be estimated by:

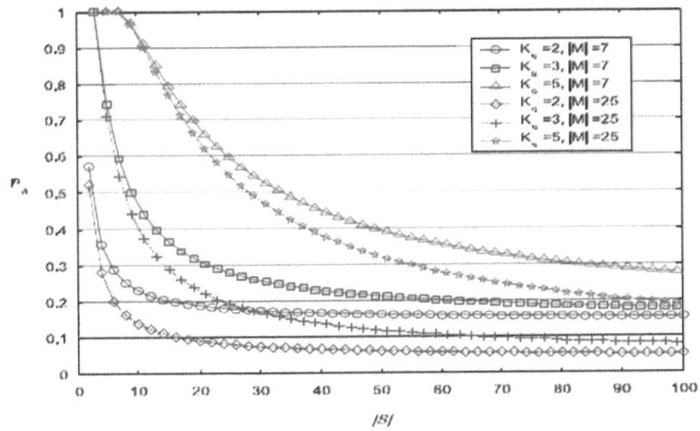

Fig. 4 Probability for key establishment between sink and access node

$$Pd = 1 - (1 - PsaPm)^{g} \cdot (1 - PmPsaPs)^{g*d} \cdot (1 - PmPaPsa)^{g(g-1)}$$

where d denotes the average number of neighbours that a sensor can contact, and g denotes the average number of stationary access nodes that the node has in its neighbourhood [15] (Fig. 4).

5 Conclusion

In this paper we proposed a three tier grid architecture for the autonomous wireless body area network. Use of two different algorithms and signature vapour leads to WBANs being less vulnerable to the attacks done by adversaries. We have further improved the security performance of the proposed scheme against stationary access node replication attack by strengthening the authentication mechanism between stationary access nodes and sensor nodes. We used the one-way hash chains algorithm in conjunction with the different algorithmic schemes in WBAN.

References

1. Ali, A., Aslam Khan, F.: Energy-efficient cluster based security mechanism for intra-WBAN and inter-WBAN for health care applications. EURASIP J. Wireless Commun. Netw. **2013**, 216 (2013). doi:10.1186/1687-1499-2013-216
2. Paradiso, R., Loraga, G., Taccini, N.: A wearable health care system based on knitted integrated sensors. Proc. IEEE Trans. Info. Technol. Biomed. **9**(3), 337–344 (2005)
3. Djenouri, D., Khelladi, L., Badacha, N.: A survey of security issues in mobile ad hoc and sensors networks. IEEE Commun. Surv. Tutorials **7**(4), 2–28 (2005)
4. Wang, Y., Attebury, G., Ramamurthy, B.: A survey of security issues in wireless sensors network. IEEE Commun. Surv. Tutorials **8**(2), 2–23 (2006)
5. Perrig, A., Szewczyk, R., Tygar, J.D., Wen, V., Culler, D.: SPINS: security protocol for sensors network. Wireless Netw. **8**(5), 521–534 (2002)

6. Venkatasubramaniam, K.K., Gupta, S.K.S.: Physiological vale-based efficient usable security solutions for body sensor networks. ACM Trans. Sens. Netw. (TOSN) **6**(4), 60–68 (2010)
7. Van Laerhoven, K., Lo, B.P.L., Ng, J.W.P., et al.: Medical healthcare monitoring with wearable and implantable sensors. In: 3rd International Workshop on Ubiquitous Computing for Pervasive Healthcare Applications (UbiHealth) [Online]. Available: http://www.healthcare.pervasive.dk/ubicomp2004/papers/final_papers/laerhoven.pdf
8. Darwish, A., Hassanien, A.F.: Wearable and implantable wireless sensor network solutions for healthcare monitoring. Sensors **11**, 5561–5595 (2011)
9. Kumar, P., Lee, H.J.: Security issues in healthcare applications using wireless medical sensor networks: a survey. Sensors **12**(1), 55–91 (2012)
10. Selimis, G., Huang, L., Masse, F., Isekoura, I., Ashouei, M., Catthoor, F., Huisken, J., Stuyt, J., Dolmans, G., Penders, J., De Groot, H.: A lightweight security scheme for wireless body area networks: design, energy evaluation and proposed microprocessor design. J. Med. Syst. **35**, 289–1298 (2011)
11. Al Ameen, M., Liu, J., Kwak, K.: Security and privacy issues in wireless sensor networks for healthcare applications. J. Med. Syst. **36**, 93–101 (2012). doi:10.1007/s10916-010-9449-4
12. Campbell, A.T., Eisenman, S.B., Lane, N.D., Miluzzo, E., Peterson, R.A., Lu, H., Zheng, X., Musolesi, M., Fodor, K., Ahn, G.: The rise of people-centric sensing. IEEE Internet Comput. **12**(4), 12–21 (2008)
13. Jovanov, E., Milenkovic, A., Otto, C., de Groen, P.C.: A wireless body area network of intelligent motion sensors for computer assisted physical rehabilitation. J. Neuro. Eng. Rehabil. **2**, 6 (2005)
14. Dohler, A.: Wireless sensor networks: the biggest cross-community design exercise to-date. Recent Patents Comput. Sci. **1**, 9–25 (2008)
15. Kouvatsos, D., Min, G., Qureshi, B.: Performance issues in a secure health monitoring wireless sensor network. In: Proceedings of 4th International Conference on Performance Modelling and Evaluation of Heterogeneous Networks (HET-NETs'2006), British Computer Society (BCS), IEE, Ilkley, UK, 11–13 Sept 2006, pp. WP01(1–6)

Ensemble Classifier for Concept Drift Data Stream

S. Senthamilarasu and M. Hemalatha

Abstract In this era, an emerging field in the data mining is data stream mining. The current research technique of the data stream is classification which mainly focuses on concept drift data. Mining drift data with the single classifier is not sufficient for classifying the data. Because of the high dimensionality and failure to get processed within considerable time and memory, false alarm rate is high, and classification accuracy result is low. In chapter, a proposed genetic-based intuitionistic fuzzy version of k-means has been introduced for grouping interdependent features. The proposed method achieves improvement in classification accuracy and perhaps in selecting the least number of features which show the way to simplification of learning task. The experiment shows that the advocated method performs well when compared with existing methods.

Keywords Intuitionistic • Fuzzy logic • K-means • Ensemble classification • Data mining

1 Introduction

One of the most well-known systems in the prescient information is data classification. For decades, most of the analysts have focused on inductive learning strategies for performing real-world applications. In single model applications, it is difficult to explain the drifting data because of infinite volume and it is continuously evolving [1]. Incorporating the new data into the model without rehashing the whole learning process is complicated, and the model is not embraced while the data is conceptually overhauled [2] [3]. Due to this reason, ensemble strategies have recently acquired a great deal of attention in the machine learning models.

S. Senthamilarasu • M. Hemalatha (✉)
Department of Computer Science, Karpagam University, Coimbatore, TN, India
e-mail: s.a6asu@gmail.com; csresearchhema@gmail.com

E.B. Rajsingh et al. (eds.), *Informatics and Communication Technologies for Societal Development*, DOI 10.1007/978-81-322-1916-3_13, © Springer India 2015

Ensemble techniques such as boosting and bagging have proven to be highly efficient in a data mining framework [4] [5].

This study utilizes genetic-based intuitionistic fuzzy, a fuzzy classifier that is part of the MATLAB. Bagging method is the ensemble approach whose accuracy is compared to that of a genetically based intuitionistic fuzzy classifier. The statistical tests were performed with the intuitionistic fuzzy classification. Here, the UCI repository data set is used for the experimental analysis. A feature selection process was applied to some of them to make them suitable for use with genetic-based intuitionistic fuzzy. In addition, a sample stratification process was performed on all data sets for class distribution preservation purposes.

2 Related Work

There are several works which classify a large amount of data classification models, ensemble learning framework has been improved from a variety of problems to handle a large amount of stream data, expensive cost, and concept drifting problems [15]. Multi-label ensemble active learning can reduce the label cost and solve the concept drift problems [14]. The objective of this algorithm minimizes the loss of classifiers using the weights in the learning methods to predict the data with high accuracy rate. Active learning is to dynamically adjust the weight of each base classifier based on its prediction accuracy of data [6]. Streaming ensemble algorithm (SEA) has presented for classifying the large drifting data, i.e., splitting the large set into different chunk sizes and executed into the classifier with the fixed size using heuristic replacement strategy [7]. Weighted ensemble classifier is used to mine the concept drifts accurately. The single classifier models are not discovered to produce good classification results against drifting data [8]. Two types of ensemble algorithms are presented for classifying uncertain data, namely, static classifier ensemble (SCE) and dynamic classifier ensemble (DCE). These algorithms performed well, but DCE for uncertain data outperforms the SCE [5]. From the review of work, the uncertainty of data stream is not able to permanently maintain an accurate model. So, the ensemble of intuitionistic ensemble classifier is introduced to improve and maintain accurate rate.

3 Feature Selection

Feature selection (FS): The purpose of FS is to find out or choose the best subset feature from all other features in the data set. Selecting possible features means discovering more information in the large database for classifying the results. Basically, the proposed method has been derived from k-means, fuzzy k-means. Data dimensionality leads to decrease classifier's performance, and increase the computational time with more features contained in it. To defeat this issue, we should reduce the features and select the more appropriate feature from among the original features.

4 Methodology

4.1 Intuitionistic Fuzzy k-Means (IFKM)

Intuitionistic fuzzy k-means clustering algorithm is based upon intuitionistic fuzzy set theory. Fuzzy set generates only membership function $\mu(x)$, $x \in X$, whereas intuitionistic fuzzy set (IFS) considers both members $\mu(x)$ and nonmembers $\nu(x)$. An intuitionistic fuzzy set A in X is written as equation (1):

$$A = \{x, \mu_A(x), \nu_A(x) | x \epsilon X\} \tag{1}$$

where $\mu_A(x) \rightarrow [0, 1], \nu_A(x) \rightarrow [0, 1]$ are the membership and nonmembership degrees of an element in set A with the condition as mentioned in equation (2):

$$0 \leq \mu(x) + \nu_A(x) \leq 1 \tag{2}$$

when μ $\mu_A(x) \rightarrow [0, 1], \nu_A(x) \rightarrow [0, 1]$ for every x in the set A, then the set A becomes a fuzzy set. All intuitionistic fuzzy sets indicate a hesitation degree, $\pi_A(x)$ which arises due to the lack of knowledge in defining the membership degree of each element, x in the set A, and is given by the equation (3):

$$\pi_A(x) = 1 - \mu_A(x) - \nu_A(x); 0 \leq \pi_A(x) \leq 1 \tag{3}$$

Due to hesitation degree, the membership values lie in the interval as mentioned in the equation (4):

$$[\mu_A(x), \mu_A(x) + \pi_A(x)] \tag{4}$$

IFKM minimizes the objective function as in equation (5):

$$J_{IFKM} = \sum_{r=1}^{K} Jr \sum_{r=1}^{k} \sum_{i=1}^{n} \mu_{ri}^{\rightarrow m} d(x_i, c_r)^2 + \sum_{i=1}^{k} \mu_{ri}^{\rightarrow} e^{1-\pi_r^*} \tag{5}$$

where

$$\mu_{ri}^{\rightarrow} = \mu_{ri} + \pi_{ri} \tag{6}$$

and where $\mu_{\vec{ri}}$ denotes the intuitionistic fuzzy membership and μ_{ri} denotes the conventional fuzzy membership of the i-th feature in r-th cluster defined as in equation (6):

$$\mu_{\downarrow} ri = 1/(\Sigma_{\downarrow}(l-1) \uparrow k\Big(\big(d(x_{\downarrow}i, c_{\downarrow}r)\big)/\big(d(x_{\downarrow}i, c_{\downarrow}l)\big)^{\uparrow}(2/((m-1)))\Big)) \tag{7}$$

In this chapter, the eigenvalue is adopted and incorporated into the distance measure conventional FCM as in equation (8):

$$d(x_i, c_r) = \frac{\|x_i - c_r\|^2}{\sigma_r} \tag{8}$$

and σ is the weighted mean distance of cluster r and is given by equation (9):

$$\sigma_r = \left\{ \frac{\sum_{i=1}^{n} \mu_{ri}^{\to m} \cdot \|x_i - c_r\|^2}{\sum_{i=1}^{n} \mu_{ri}^{\to m}} \right\} \tag{9}$$

π_{ri} is the hesitation degree, which is defined as follows in equation (10):

$$\pi_{ri} = 1 - \mu_{ri} - \left(1 - \mu_{ri}^{\alpha}\right)^{1/\alpha}, \quad \alpha > 0 \tag{10}$$

and is calculated from Yager's intuitionistic fuzzy complement; the intuitionistic fuzzy set becomes (11)

$$A^{IFS} = \left\{x, \mu_A(x), \left(1 - \mu_A(x)^{\alpha}\right)^{1/\alpha | x \epsilon X\}} \tag{11}$$

$$\pi_{\vec{r}} = \frac{1}{N} \sum_{i=1}^{n} \pi_{ri} \tag{12}$$

The eigen value against scaled co-ordinate index of the attribute of the data values are depicted in the Fig. 3. For every feature with distinct membership and in deterministic values assigned to k diverse groups. On the ground that with intuitionistic fuzzy every feature may not fit in only one group and it is much better to every feature of correspondence features in whole group. Consequently, throughout the choice of the best feature, more accurate relations between genes are accessible [11]. The characteristic feature is not correlated with other members of one particular cluster; it corresponds to the entire cluster to be chosen with crisp attribute techniques. For clustering characteristics, the intuitionistic k-modes algorithm is utilized in which the mode of every cluster is characterized as one of its characteristics is having the largest different association excess measure around the different offers in that group. An altered form of (6), where distance function is characterized as the reverse of interdependency measure, the equation is given below in (13):

$$\mu_{\vec{ri}} = \frac{1}{\sum_{i=1}^{k} \left[\frac{R(A_i, n_l)}{R(A_i, n_r)}\right]^{2/_{m-1}}} + \left[1 - \frac{1}{\sum_{i=1}^{k} \left[\frac{R(A_i, n_l)}{R(A_i, n_r)}\right]^{2/_{m-1}}} - \left[1 - \left[\frac{1}{\sum_{i=1}^{k} \left[\frac{R(A_i, n_l)}{R(A_i, n_r)}\right]^{2/_{m-1}}}\right]^{\alpha}\right]^{1/_{\alpha}}\right] \tag{13}$$

where li is the number of clusters, $\mu_{\vec{ri}}$ is intuitionistic fuzzy membership degree of i-th feature in r-th cluster, and m is a weighting exponent. Afterwards, to satisfy the condition (4) $\mu_{\vec{ri}}$ is normalized by Eq. (14):

$$\mu_{\vec{ri}} = \frac{\mu_{ri}^{\to new}}{\sum_{i=1}^{k} \mu_{li}^{\to old}} \tag{14}$$

In this approach, the mode of a cluster is upgraded to the feature with the most elevated fuzzy multiple interdependence redundancy in that cluster. Fuzzy multiple interdependence redundancy of every characteristic feature ought to be ascertained, calculated regardless of its own membership degree in the associated cluster. Acknowledging this intuitionistic fuzzy membership degree, the mode of every cluster will never show signs of change, since it has high membership degree in that cluster.

4.2 Novel Genetic Algorithm-Based Intuitionistic Fuzzification Techniques in Selection of Features from Clusters

After clustering of features, the best one ought to be chosen in every group. In this chapter genetic algorithm method is proposed with a specific end goal to consider intuitionistic fuzzy membership of features in cluster throughout determination [9]. At the outset, every feature is thought to fit in with a group which has the most astounding membership degree [10]. Members of r-th cluster are represented by member r. The rank of each feature in each cluster is calculated and the one with the highest ranks are selected as the candidates. In calculating the rank of each feature in intuitionistic fuzzy approach, the technique involves assigning a positive weight to the interdependent features which belong to the members of cluster r. It also assigns a negative weight as a penalty to the interdependent features of nonmembers the cluster r. The assigned weight is equal to the membership degree of the feature in its cluster. Therefore, the selected feature would be more interdependent with its co-cluster features and less interdependent with other clusters' members. Calculating the rank of each feature in this method is performed according to (15):

$$rank_r(A_i) = \mu_{ri}^{\rightarrow m}\left[\sum_{j\in members_r, j\neq i} \mu_{rj}^{\rightarrow m} R(A_i : A_j)\right] - \mu_{ri}^{\rightarrow m}\left[\sum_{j\in members_r, j\neq i} \mu_{rj}^{\rightarrow m} R(A_i : A_j)\right] \quad (15)$$

It is considered as the fitness function of the genetic algorithm. The features with the highest fitness rank are selected as the best candidates to perform the crossover operation. The genetic algorithm uses in addition the insertion operator at particular iterations to overcome the problem of local optima. So, it adapts the concept of genetic algorithm restarting in which it deletes some of the instances from the current population and adds the same amount of instances from the initial population which is left during the initial stage of random population generation. Evaluation of feature ranking technique is considered as the fitness function. The candidates with the highest ranking is considered as holding the highest fitness value, and they are taken into consideration for further processing in genetic algorithm restarting.

5 Result and Discussion

In this section, the proposed method advocates uncertain data stream classification, i.e., concept drift and simulating the uncertain data in nontemporal and temporal data set. Here, temporal and nontemporal data sets are used with categorical attributes in our experiments. The proposed framework facilitates anytime learning on problems of arbitrary size, and this method is efficient, as accurate as a single classifier, and adjusts quickly to changes in the target. Moreover, the method is easy to implement and independent of the underlying classifier. When comparing the results based on accuracy in single classifiers, use of ensemble classification provides significantly better performance [13]. One direction is to examine the diversity mechanism, i.e., different blocks of data for the various classifiers to create classification diversity, which is necessary for effective ensembles, and to experiment with other approaches in promoting diversity, such as the use of different classification methods. Limited data are sampled and different methods are combined to predict the targets. Another simple variation is to keep a working set of high-quality classifiers that are slightly larger than the ensemble size and test

Table 1 Analysis of ensemble classifier using bagging methods

Data set	No bag	Bags with majority vote	Bags with intuitionistic fuzzy
Sensor	96.51	97.07	97.63
Electricity	97.33	97.89	98.45
Breast cancer	97.19	97.75	98.31
Iris	97.73	97.30	96.93
Diabetes	96.55	96.71	97.10
Heart disease	97.33	97.33	98.31

all the combinations of evaluation set. This "best *n-of-m*" strategy moves us somewhat closer to a globally optimal set while limiting the additional computation. In order to properly "optimize" an ensemble, it is important to have a clear understanding of the mechanisms through which ensembles achieve their effectiveness. In a related work in the direct optimization of ensembles via evolutionary search, we are in the early stages of a data mining task over the space of possible ensembles, varying ensemble characteristics and allowing a group of ensembles to compete for limited resources based on accuracy.

In order to achieve a higher classification accuracy by considering data uncertainty, we might as well execute average uncertainty decision tree and apply them to 6 real data sets (see tables) taken from the UCI Machine Learning Repository. The amount of groups discovered by GA-based intuitionistic FCM was taken as the random partitions. In the clustering part examination, a simple majority vote is utilized to join classification predictions. In the clustering experimental tests, an expert combination could be utilized, i.e., FCM grouping features via centroids. For test case, the closest centroid is resolved based on the group centroid quality. This centroid compares to a cluster of training data, from which a decision tree was made. The decision tree is consulted for a classification prediction. The effects of training and testing ensemble of classifiers are consistent with a few partitioning techniques depicted above and could be seen in Table 1. Figure 1 presents the feature index of important attribute selection. For selecting attribute, threshold value is used to distinguish ranges of values where the behavior predicted by the model varies in some important way [12]. Figure 2 represents the mean square error rate of a number of growing trees based on bagging methods.

Table 2 presents the accuracy results of the proposed method and other existing ensemble classification methods. The proposed method seems to more consistent than the other existing methods. Table 3 represents the proposed method of intuitionistic fuzzy classifier ensemble compared with single classifiers such as SVM, C4.5, and CART. Based on the accuracy case, this suggested method is gaining much better result than the other methods. Figure 4 represents the true positive rate (TPR) versus false positive rate (FPR) of the ensemble classifier of the sensor data set. The rate of plotting probability distributions for detection and false alarm is known for both.

So far investigated is the performance of intuitionistic fuzzy ensemble classifier (IFEC) method with other existing methods such as support vector machine (SVM),

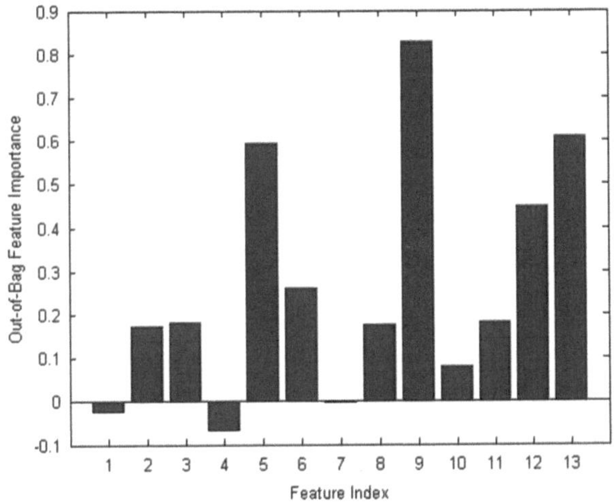

Fig. 1 Feature selection index

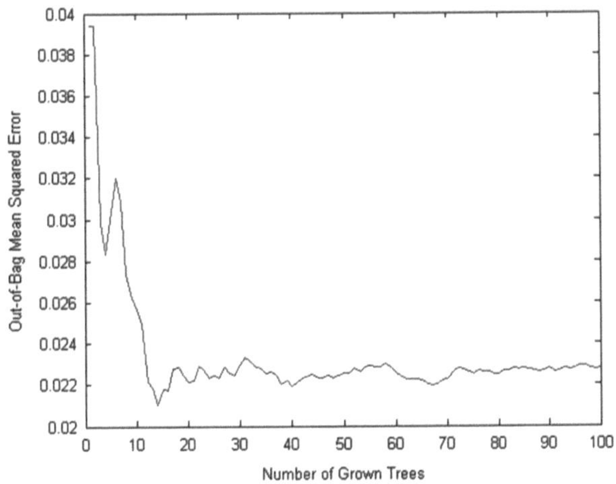

Fig. 2 Mean square error rate based on growing trees

C4.5, and CART for concept summarization. Existing methods and IFEC both perform data chunking into clusters. However, IFEC method converts each data chunk into a virtual sample, while support vector-based clustering (SVBC) utilizes the support vectors of one-class classifier to represent the chunked data. In terms of the number of k in k-means clustering, it remains the same as it assumes the number of known concepts of the users.

The experimental result discovers IFEC that performs significantly better than SVBC. IFEC method converts each data chunk into a virtual sample by transforming the original feature values in each chunk into some histogram format and thus it loses much data information of the chunked data. However, SVBC utilizes the support vectors of one-class classifier, which can describe the data distribution of the target class, to represent the chunked data. As a result, IFEC

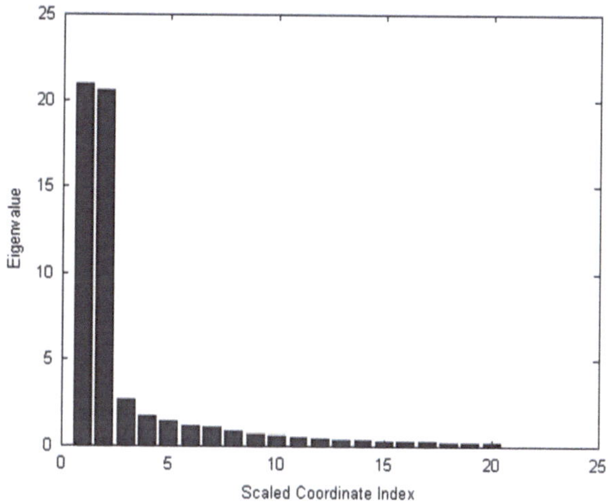

Fig. 3 Eigenvalue measure against scaled coordinate index

Table 2 Analysis of ensemble classifier compared with other ensemble classifiers

Data set	Naïve Bayes	Multi-label data	Fuzzy classifier	Intuitionistic fuzzy classifier
Sensor	95.60	96.83	97.14	97.36
Electricity	95.87	96.56	97.77	98.45
Breast cancer	95.53	96.67	97.89	98.52
Iris	94.93	95.31	96.18	96.93
Diabetes	96.31	96.58	96.89	97.10
Heart disease	95.97	97.15	97.63	98.31

Table 3 Analysis of ensemble classifier compared with other single classifiers

Data set	SVM	C4.5	CART	Intuitionistic fuzzy ensemble classifier
Sensor	89.65	92.62	93.64	97.36
Electricity	90.42	92.00	92.36	98.45
Breast cancer	94.26	96.48	96.71	98.52
Iris	93.63	95.81	96.52	96.93
Diabetes	94.41	95.02	95.98	97.10
Heart disease	93.97	95.71	96.36	98.31

significantly outperforms SVBC method in concept summarization learning. For each data stream, the prediction accuracy for each concept is also illustrated. It is noted that the SVBC method always outperforms IFEC method for each concept prediction.

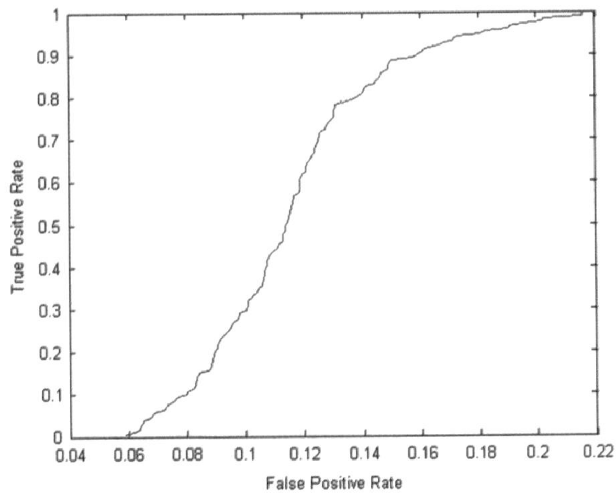

Fig. 4 True positive vs. false positive rate

6 Conclusion

In this chapter, a novel approach intuitionistic fuzzy k-means clustering is presented. The results presented in this chapter suggest that for very large data sets, the creation of ensembles of classifiers can perform reasonably well. Interestingly, the results indicate that bagging of individual partitions can yield better results than learning from the entire data set. It is surprising as bagged classifiers created from subsets, in effect, carry much less data. This proposed method shows more consistent performance than the other existing stand-alone methods, i.e., 2.5 % than CART, 3.09 % than C4.5, and 5.20 % than SVM. It also outperforms other ensembles: 0.73 % higher than fuzzy classifier, 1.51 % than multi-label data, and 2.48 % than the naïve Bayes classifier. The experimental results show that the new ensemble of decision trees outperforms state-of-art stand-alone and ensemble methods.

References

1. Wenhua Xu, Zheng Qin, Yang Chang: A framework for classifying uncertain and evolving data stream. Inf. Technol. J. **10**(10), 1926–1933 (2011)
2. Biao Qin, Yuni Xia, Sunil Prabhakar, Yicheng Tu: A rule-based classification algorithm for uncertain data. In Proceedings IEEE International Conference on Data Engineering, pp. 1666–1640, Shanghai, China (2009)
3. Shirui Pan, Kuan Wu, Yang Zhang, Xue Li: Ensembles of fuzzy classifiers. In: Proceedings International Conference on Advances in Knowledge Discovery and Data Mining. Lecture Notes in Computer Science, vol. 6118, pp. 488–495, Phuket, Thailand (2010)
4. Ai-Min Yang, Ling-Min Jiang, Xin-Guang Li, Yong-Mei Zhou: A novel fuzzy classifier ensemble system. In: Proceedings IEEE International Conference on Machine Learning and Cybernetics, vol. 6, pp. 3582–3587, Hongkong, China (2007)
5. Canul Reich, J., Shoemaker, L., Hall, L.O.: Ensembles of fuzzy classifiers. In: Proceedings IEEE International Conference on Fuzzy Systems, pp. 1–6, London, UK (2007)

6. Peng Wang, Peng Zhang, Li Guo: Mining multi-label data streams using ensemble-based active learning. In: Proceedings of SIAM SDM 2012, pp. 1131–1140, California, USA (2012)
7. Street, W.N., Kim, Y.S.: A streaming ensemble algorithm (SEA) for large scale classification. In: Proceedings of the 7th International Conference on Knowledge Discovery and Data mining, pp. 377–382, San Francisco, CA, USA (2001)
8. Haixun Wang, Wei Fan, Philip S. Yu, Jiawei Han: Mining concept-drifting data streams using ensemble classifiers. In: Proceeding of the 9th ACM SIGKDD International Conference on Knowledge Discovery and Data Mining (KDD'03), pp. 226–235, Washington DC, USA (2003)
9. Duangsoithong, R., Windeatt, T.: Bootstrap feature selection for ensemble classifiers. In: Advances in Data Mining Applications and Theoretical Aspects. Lecture Notes in Computer Science, vol. 6171, pp. 28–41. (2010)
10. Senthamilarasu, S., Hemalatha, M.: A genetic algorithm based intuitionistic fuzzification technique for attribute selection. Indian J. Sci. Technol. **6**(4), 4336–4346 (2012)
11. Ching Wei Wang: New ensemble machine learning method for classification and prediction on gene expression data. In: Proceeding of the 28th IEEE EMBS Annual International Confer ence, New York City, USA, Aug 30–Sept-3, pp. 3478–3481. (2006)
12. Bauer, E., Kohavi, R.: An empirical comparison of voting classification algorithms: bagging, boosting, and variants. Mach. Learn. **36**(1–2), 105–139 (1999)
13. Rokach, L.: Ensemble based classifiers. Artif. Intell. Rev. **33**, 1–39 (2010)
14. Nikulin, V., McLachlan, G.J., Ng, S.K.: Ensemble approach for the classification of imbalanced data. In: Proceedings of the 22nd Australasian Joint Conference on Advances in Artificial Intelligence (AI'09), pp. 291–300. (2009)
15. Shirui Pan, Yang Zhang, Xue Li: Dynamic classifier ensemble for positive unlabeled text stream classification. Knowl. Info. Syst. **33**(2), 267–287 (2012)

Automatic Adaptation of Frames Using NSEW Affine Translation

D. Raveena Judie Dolly, G. Josemin Bala, and J. Dinesh Peter

Abstract Video compression standards follow certain principles for frame identification based on group of pictures (GOP). It was observed that instead of following the precoded GOP, if frames are adaptively chosen based on simple matching criteria, further processing could be effectively done. It was also noted from the earlier techniques that the presence of B-frames leads to buffering of the memory within the past and future frames during the coding stage. The computational time increases with more number of B-frames. Hence, a novel framework has been emphasized in the present investigation firstly for adaptive frame identification and secondly for replacing B-frames with either I-frames or P-frames using NSEW affine translation for the purpose of video compression. The result shows that the performance objective evaluation coincides with the subjective evaluation.

Keywords Frame identification • Group of pictures • I-frame • P-frame • B-frame • Video compression • Adaptive GOP

1 Introduction

The conventional video coding standards use intra ("I") and inter ("P (prediction)" and "B (bidirectional)") frame methods for improved video coding. The frames are coded as a group of pictures (GOP) comprising one intra frame with subsequent inter frames. To maximize the coding efficiency, it is important to determine the distribution of the three frame types, also called the GOP structure. In [1], the GOP structure is determined dynamically in real time by using the information such as

D.R.J. Dolly (✉) • G.J. Bala
Department of ECE, Karunya University, Coimbatore, TN, India
e-mail: dollydinesh@karunya.edu

J.D. Peter
Department of IT, Karunya University, Coimbatore, TN, India

E.B. Rajsingh et al. (eds.), *Informatics and Communication Technologies for Societal Development*, DOI 10.1007/978-81-322-1916-3_14,

the sum of absolute difference and the mean of absolute difference. Generally, in conventional techniques the key reference frame is the I-frame. Paul et al. [2] insist on multiple reference frames (MRFs), that is, the I-frames which are considered to improve the coding performance of the H.264 video coding standard over other contemporary coding standards. MRFs facilitate better predictions than using just one reference frame, for video with repetitive motion, uncovered background, lighting change, etc. In general, if a sequence does not contain any scene changes or extremely high motion activity compared to the previous frames, insertion of I-frames reduces the coding performance. Therefore, we need to insert optimal I-frames [2]. But if scene change is detected on some frames, the frame coding types of those frames have to be changed to "I" type [3]. When another I-frame is inserted for the next GOP, better image quality (with the cost of more bits) is recovered and then again quality degrades toward the end of GOP. As a result, the further an inter frame is away from the I-frame, the lower the quality becomes [2]. B-frame coding needs reference frames from the future frames; therefore, the future frame has to be encoded before the encoding of B-frame which adds to delay and in turn takes more computation time [4]. Also, a new Gestalt factor called "shift and hold" is presented for 2D object extraction algorithms in [5], where shifting is dealt in detail which motivated the authors in performing shifting and then correlation for replacement of B-frames. In [6], it is said that the motion model is a 2D transformation (affine transformation or homography) of an image of the object (e.g., the initial frame). In [7], traditional motion-compensated coders use only translational motion estimates for motion-compensated prediction, but the affine motion model is considerably more expressive, because it incorporates transformations such as rotation and shear. It allows higher adaptability to the underlying image, and it is expected that it will yield better results [8]. The author suggests affine transformation for better results. So, an affine translation is adopted for replacement of B-frames. The proposed methodology deals with adaptive frame identification where the importance of adaptive frame identification is emphasized. The first frame is considered to be the I-Frame. The similarity between the adjacent frames is performed and identified as either P-frames or B-frames. If scene change occurs, the frame is identified as I-frame. The matching criteria are performed using correlation coefficient, where A and B are matrices or vectors of the same size.

$$R = \frac{\sum_m \sum_n (A_{mn} - \overline{A})(I_{mn} - \overline{I})}{\sqrt{\left(\sum_m \sum_n (A_{mn} - \overline{A})^2\right)\left(\sum_m \sum_n (I_{mn} - \overline{I})^2\right)}} \tag{1}$$

Where $\bar{A} = \text{mean}(A)$, and $\bar{I} = \text{mean}(I)$.

This chapter is organized as follows. Section 2 explains in detail about the adaptive frame identification and determination of I-frames, B-frames, and P-frames. Section 3 describes about the replacement of the conventional B-frame with the I-frames or P-frames (NSEW affine translation) toward better video coding. Section 4 discusses about the analysis regarding still camera|still background and still camera|moving background. Section 5 concludes this chapter.

2 Adaptive Frame Identification

The primary purpose is to compress video data without degrading the image quality. So, the I-, B-, and P-frames are adaptively identified based on scene changes. For performing the process, a video file is taken as input and is converted into frames as shown in Fig. 1. A simple matching criterion, correlation coefficient, is employed for finding the correlation between adjacent frames. If two frames are exactly similar, the correlation factor would be “1.” Based on the correlation factor, certain threshold values are fixed as $ɱ_1 = 0.7$, $ɱ_2 = 0.9$, $ɱ_3 = 0.65$, and $ɱ_4 = 0.85$ after subjective evaluations by various experts and many experimentations.

The determination of I-frame, P-frame, and B-frame is identified based on the correlation factor.

2.1 *Algorithm for Adaptive Frame Identification*

Step 1: Input a video file.
Step 2: Convert the video into “N_i,” $i = 1, 2, \ldots, N$ no. of frames.
Step 3: The first frame (“N_1”) is taken as the reference frame and name it as I-frame.
Step 4: Correlation between N_i and $N_{(i+1)} \rightarrow$ (“j”) is found.
Step 5: If ($j < ɱ_1$), $N_{(i+1)}$ is stored as I-frame and “i” is incremented.
Step 6: If ($j >= ɱ_2$), $N_{(i+1)}$ is stored as P-frame and “i” is incremented.
Step 7: If ($j >= ɱ_1$ and $j < ɱ_2$), then correlation between $N_{(i+1)}$ and $N_{(i+2)} \rightarrow$ (“k”) is performed.
Step 8: If $k > ɱ_2$, $N_{(i+1)}$ is stored as B-frame.
Step 9: If $k < ɱ_2$, $N_{(i+1)}$ is stored as I-frame.

The algorithm discusses about the determination of each frame which is discussed in detail.

2.2 *Determination of I-Frame*

The first frame of a video is considered to be the reference frame for the future frames, and the same is taken to be the I-frame. Similarly when there is a scene change in the future frames, the correlation factor between the frames taken for comparison will be less. If the value is less than the fixed threshold of $ɱ_1$, it is considered to be an I-frame. It is chosen with the concept of two frames having less

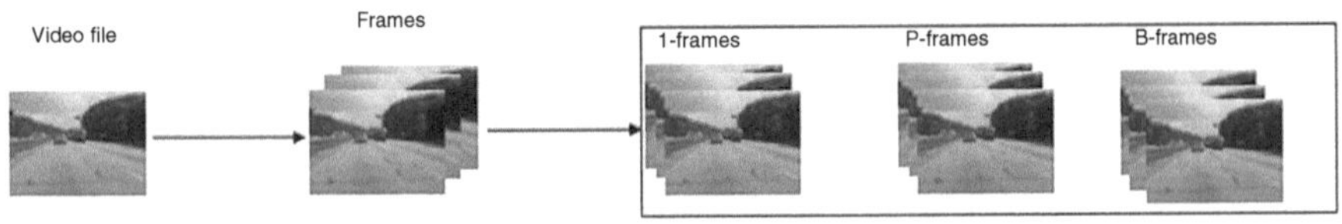

Fig. 1 Frame identification

similarity between two adjacent frames since it contains all the necessary data to recreate a complete image.

2.3 *Determination of P-Frame*

The frame is identified to be a P-frame in such a way that the similarity between frames taken for comparison will be high, that is, when the correlation between the reference and future frame exceeds the fixed threshold of η_2. Similarly the process continues by considering the P-frame as the present frame and the immediate next frame as the future frame after which the correlation is done for the present and the future frame, and it will be considered as P-frame if it exceeds a fixed threshold of $\eta_{2.}$ If not, the process continues to identify for its suitable frame.

2.4 *Determination of B-Frame*

The B-frame is identified in such a way that similarity between the past and the present frame is taken, and if there is a new object that has been introduced very lightly in the present frame, the similarity between them becomes less. In such a case the similarity between the present frame and future frame is considered since in the future frame, information about the newly introduced object will be present. So, firstly, the correlation between the present frame and the past frame is found, and if the correlation factors lie between η_1 and η_2, then correlation factor between the present frame and the future frame is found and if the correlation factor is greater than η_3, then the present frame is considered to be the B-frame, or if the similarity between the past and the present frames lies between η_1 and η_2 and simultaneously if the similarity between the present and future frames is less than η3, then the present frame is considered to be the I-frame since similarity between the present frame and the future frame is less. After which the process continues until it identifies its suitable frame.

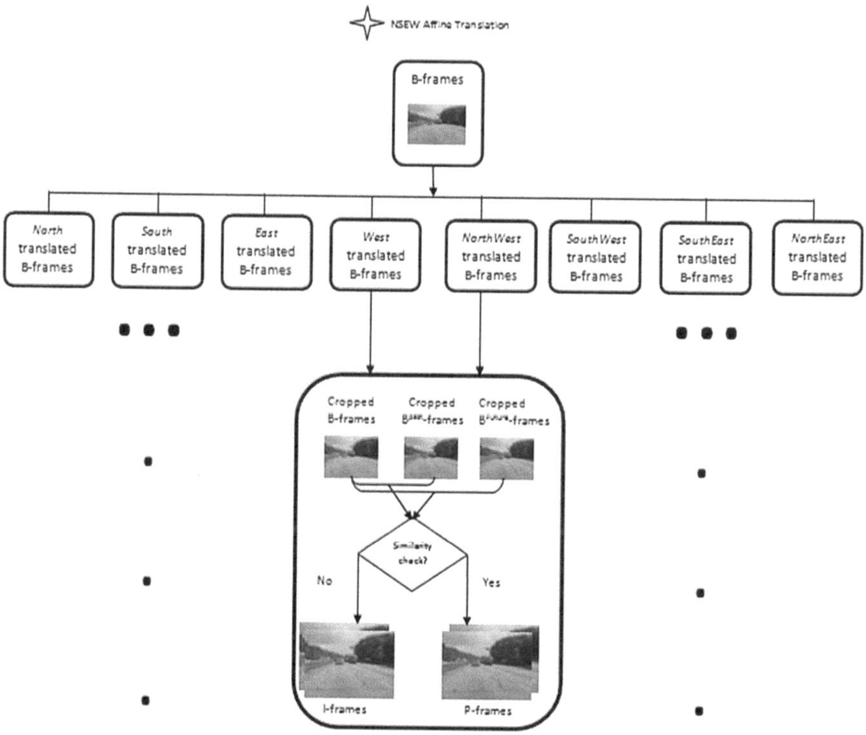

Fig. 2 NSEW affine translation

3 NSEW Affine Translation

The fundamental aim of the investigation is to replace the B-frames with either I-frames or P-frames so as to overcome the buffering of B-frames between the past and future frames while performing video compression which takes more computational time. So, a NSEW affine translation is proposed to replace the B-frames with either I-frames or P-frames (Fig. 2).

The proposed method is considered to be automatic once B-frames are encountered. All the B-frames are chosen for the purpose of conversion. Since B-frames are bidirectional, the present frame might have some features from the past and future frames. So, if translation shift is performed and cropped by the same level for the past, present, and future frames, the correlation factor between the frames might increase. If not, then the present frame is taken to be the I-frame. Once B-frames are encountered in the process, the algorithm undergoes various translations like north side(N), south side(S), east side(E), west side(W), northwest, southwest, northeast, and southeast by using the affine translation by a factor of 10 for the past, present, and future frames. The general affine transformation is commonly written in homogeneous coordinates as shown below [9].

$$\begin{vmatrix} X_2 \\ Y_2 \end{vmatrix} = A \times \begin{vmatrix} X_1 \\ Y_1 \end{vmatrix} + B \tag{2}$$

By defining only the B matrix, this transformation can carry out pure translation

$$A = \begin{vmatrix} 1 & 0 \\ 0 & 1 \end{vmatrix} \quad B = \begin{vmatrix} b_1 \\ b_2 \end{vmatrix} \tag{3}$$

After all these translations are performed, image cropping is done for all the B-frames, B^{Past} and B^{Next} frames, by the same level. The B^{Past} and B^{Next} frames can be an I- or P-frame. Matching criteria using correlation technique are performed after translation and cropping. If the correlation factor exceeds the fixed threshold of ɱ$_2$, the B-frame is considered as the P-frame since the past and future frames are much similar. If the correlation factor is found to be less than the fixed threshold of ɱ$_2$, the B-frame is considered as the I-frame since the similarity between two frames is less. The thresholds are fixed after subjective evaluations, and it was observed that the objective and subjective evaluations are almost similar.

4 Implementation

Certain observations were noted and discussed in the following sections.

4.1 *Adaptive Frame Identification*

The choices of frames chosen adaptively are discussed in detail below. The input video files are taken from [10]. All the frames are sorted as I-, B-, P-frames depending on the matching criteria [11, 12]. Based on the matching criteria discussed, the frames are identified and are shown in Table 1. The analysis is made for a still cameralstill background and a still cameralmoving background and after implementing the following number of frames was identified.

From the above table, it is seen that for a still cameralstill background, the similarity between adjacent frames is high. So, there are no I-frames or B-frames detected. So there is no need for processing B-frames which leads to buffering if the above said video file is chosen. For a still cameralmoving background, certain B-frames are noted and I-frames are also noted in vipmen.avi since many cars keep going. Even though camera is still, many new objects are introduced often which leads to many I-frames. There might be consecutive I-frames that have been observed due to the introduction of new object and its fast movement. Subjective and objective evaluations coincide to around 97 %.

Table 1 Adaptive frame identification

#	Video type	Video file	Number of I-, B-, P-frames – objective evaluation	I-, B-, P-frame numbers – subjective evaluation
1.	Still cameralstill background	Mother, Container, Akiyo	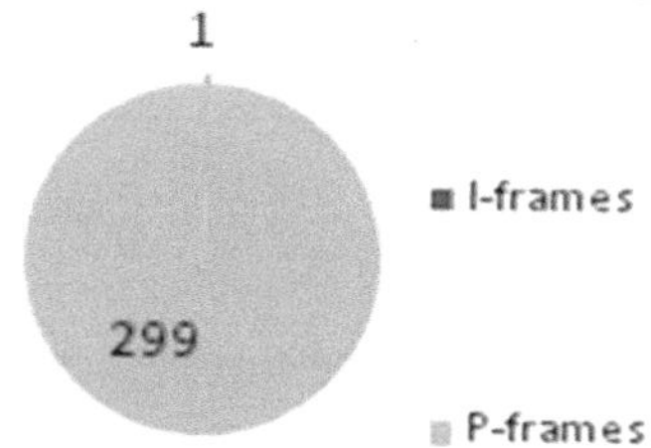	I-frame – 1 P-frames – 2 to 300 Total frames – 300
		Silent	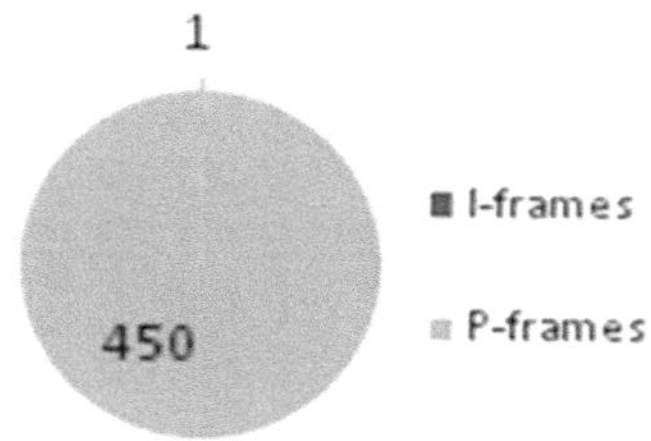	I-frame – 1 P-frames – 2 to 451 Total frames – 451
		Vipmen	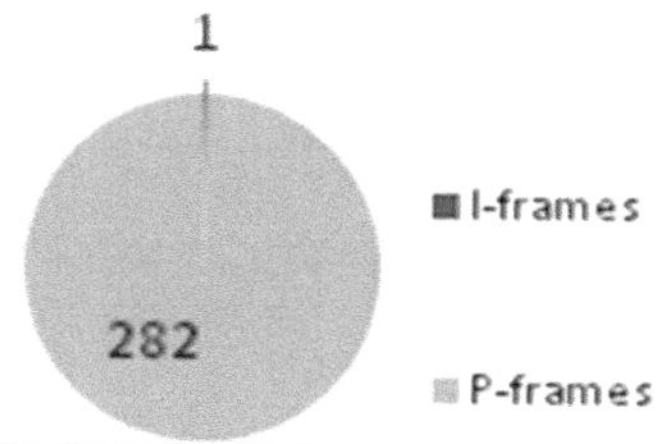	I-frame – 1 P-frames – 2 to 283 Total frames – 283

(continued)

Table 1 (continued)

#	Video type	Video file	Number of I-, B-, P-frames – objective evaluation	I-, B-, P-frame numbers – subjective evaluation
		Hall	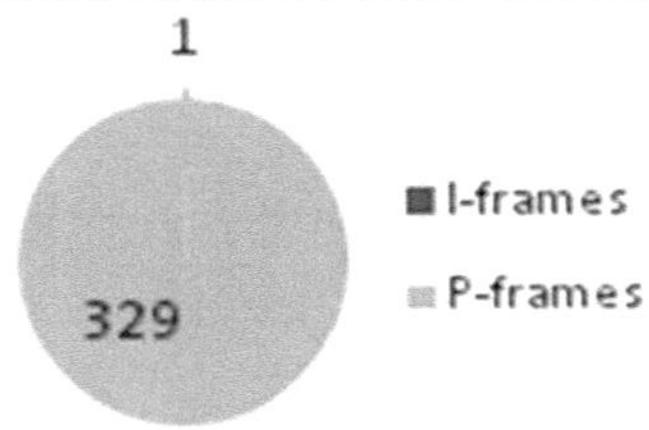	I-frame – 1 P-frames – 2 to 330 Total frames – 330
2.	Still cameralmoving background	Viptraffic	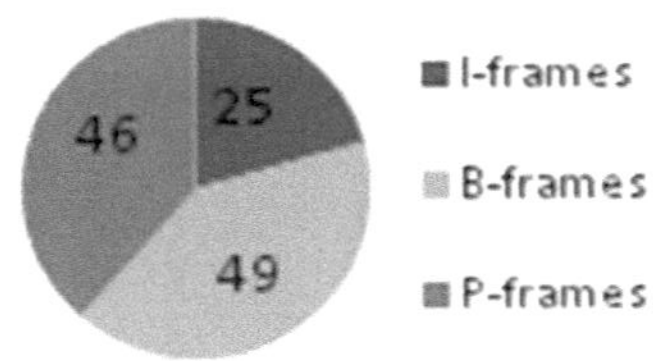	I-frames – 1, 22–24, 37–43,75–80, 93–96, 98–101 P-frames – 2–17, 29, 47, 48, 55–64, 83–87, 105–113, 118–120 B-frames – 18–21, 25–28, 30–36, 44–46, 49–54, 65–74, 81, 82, 88–92, 97, 102–104, 114–117 Total frames – 120
		News	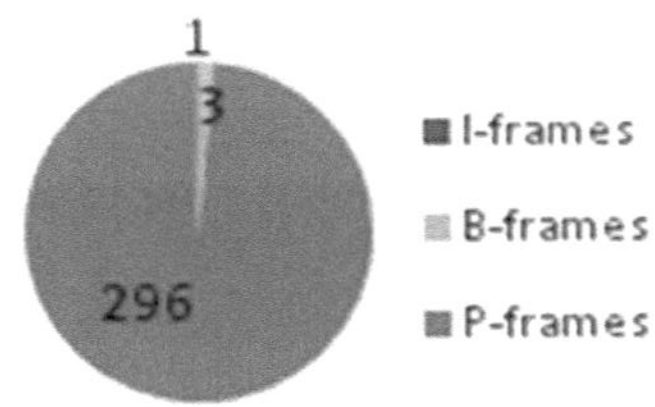	I-frame – 1 P-frames – 2–90, 92–150, 152–240, 242–300 B-frames – 91, 151, 241 Total frames – 300

Table 2 Computation time

#	Video file	Time consumed with B-frames	Time consumed without B-frames
1.	Rhinos.avi	60.530 s	40.638910 s
2.	vipmosaicking	56.958 s	41.193472 s

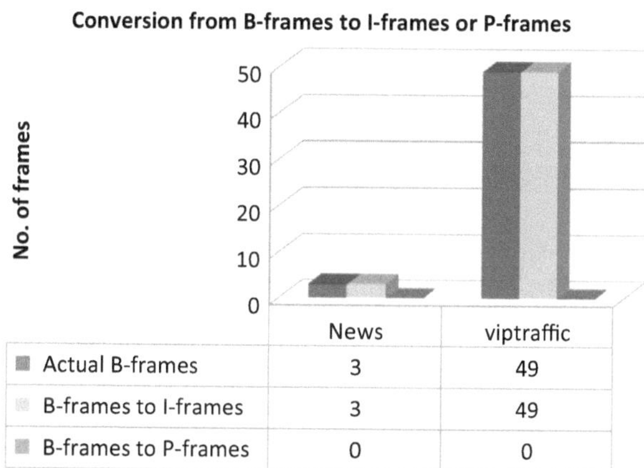

Fig. 3 Conversion of B-frames to I- or P-frames (still cameralmoving background)

4.2 *Elapsed Time with and Without B-Frames*

The elapsed time is noted by implementing the video compression using MPEG in MATLAB and is discussed in Table 2. The implementation is checked with B-frames and was found to take much time [13]. Then the B-frames are taken as either I- or P-frames depending on the input video file. The computation time taken when no B-frames are considered was less when compared. This motivated the authors to go in for replacement of B-frames with either P- or I-frames.

4.3 *NSEW Affine Translation*

The B-frames are replaced with either I-frames or P-frames to avoid buffering of memory within the past, present, and future frames. A technique, namely, NSEW affine translation, is adopted for this conversion. Successfully depending on the matching factor between frames, conversion from B to either I or P is done and is shown in Fig. 3. Subjective evaluation coincides to around 97 % with the objective evaluation.

In the above figure, the B-frames are replaced with I-frames. They are not converted to P-frames since these two video files are taken from a still camera but the background keeps moving. So, a new object is introduced often. If the camera is also moving, there may be many B-frames taking the position of P-frames. In news.avi, the background keeps moving but a still camera is in place. In Viptraffic.avi, the camera is still but many cars keep going. The above frames are considered to be B-frames which were again converted to I-frames after NSEW affine translation. If there are many I-frames, as explained in [2] the coding efficiency can be improved in terms of quality since there will be many reference frames.

5 Conclusion

The frames are adaptively identified as I-, B-, and P-frames. The B-frames are again replaced with either I-frame or P-frame. It is found that in B-frames, after NSEW affine translation, if a new object is introduced, it is converted to I-frames or if the adjacent frames are found similar, the B-frames are taken to be P-frames. The converted frames are subjectively checked and verified and found that subjective and objective evaluation coincides to around 97 %.

References

1. Yu-Lin Wang, Jing-Xin Wang, Yen-Wen Lai, Alvin, W.Y. Su: Dynamic GOP structure determination for real-time Mpeg-4 advanced simple profile video encoder. IEEE Trans. 0-7803-9332-5/05 (2005)
2. Paul, M., Lin, W., Lau, C.T.: Video coding with dynamic background. Springer, EURASIP J. Adv. Signal Process. **11**, 1–17 (2013)
3. Jens-Rainer Ohm, *Member, IEEE,* Gary J. Sullivan, *Fellow, IEEE,* Heiko Schwarz: Comparison of the coding efficiency of video coding standards—including High Efficiency Video Coding (HEVC). IEEE Trans. Circ. Syst. Video Technol. **22**(12), 1669–1684 (2012)
4. Texas Instruments-Application report. Overview of temporal scalability with scalable video coding (SVC), SPRABG3 (2010)
5. Lu, Y., Ze-Nian, L.: Automatic object extraction and reconstruction in active video. Pattern Recognit. **41**, 1159–1172 (2008)
6. http://en.wikipedia.org/wiki/Video_tracking
7. Wiegand, T., *Member, IEEE*, Steinbach, E.: Affine multipicture motion-compensated prediction. IEEE Trans. Circuits Syst. Video Technol. **15**(2) (2005)
8. Tabatabai, A.J., Jasinschi, R.S., Naveen, T.: Motion estimation methods for video compression – a review. Elsevier- J. Franklin Inst. **335B**(8), 1411–1441 (1998)
9. http://homepages.inf.ed.ac.uk/rbf/HIPR2/affine.htm
10. http://meru.cecs.missouri.edu/free_download/videos
11. Kan-Li Huang, Yi-Shin Tung et al.: A frame based MPEG characteristics extraction tool and its application in video transcoding. IEEE Trans. Consum. Electron. **48**(3), 522–531 (2002)
12. Takaaki Hamamoto, Tian Song, Takafumi Katayama, Takashi Shimamoto: Complexity reduction algorithm for hierarchical B-Picture of H.264/Svc. Int. J. Innovative Comput. Inform. Control **7**(1), 445–457 (2011)
13. Xiangyang Ji, Debin Zhao et al.: B-picture coding in AVS video compression standard. Elsevier-Signal Process: Image Commun. (2008)

Tracking of Unique Colored Objects: A Simple, Fast Visual Object Detection and Tracking Technique

Emmanuel Joy and J. Dinesh Peter

Abstract Detection and tracking of moving objects in video scenes is the important initial step for extraction of information in many computer vision applications. This idea can be used in video surveillance; industrial quality testing; traffic navigation in the forms of driver assistance, speed monitoring, etc.; human-computer interaction; sports analysis; machinery quality testing; and also in the medical field. Many simple vision problems require fast and accurate tracking. Traditional tracking methods have the disadvantage of too much time consumed by data process, and an object gets undetected under complex background. Several advanced tracking algorithms are in existence, each of them having their own positives and limitations. Nevertheless, many higher-end algorithms also fail for purposes of tracking in simple applications, because of their complexity in performance, impractical time management, and extensive memory usage. In this paper, we are discussing a simple visual tracking algorithm to track and differentiate different colored objects.

Keywords Visual tracking • Region of interest • Threshold • Moments • Saturation points • Gaussian kernel

1 Introduction

Visual tracking is an important aspect of computer vision and it has many applications. It is even now an open problem owing to the increasing complex levels of objects and classes. Indeed, it is really challenging to manage or cope with the loss of information of a 3D world on a 2D image obtained from a video frame. It is difficult to obtain the same level of details present on a real-world 3D on a 2D image.

E. Joy (✉) • J.D. Peter
Department of Information Technology, Karunya University, Coimbatore, TN, India
e-mail: emmjoy@live.in

E.B. Rajsingh et al. (eds.), *Informatics and Communication Technologies for Societal Development*, DOI 10.1007/978-81-322-1916-3_15, © Springer India 2015

It is also highly impractical to use 3D videos in every aspect of tracking because of memory management, compatibility, computational efforts, and performance issues [1–3]. Therefore, algorithm developers have concentrated more on developing methods for 2D tracking except for advanced applications that inevitably need 3D tracking schemes. Other challenges and problems can be done by noise in images, excessive disorder of background, sudden object motion changes in complex directions, short-term or long-term occlusions, low-light conditions or unexpected lighting changes, and also real-time free capture and free flow requirements. The articulated nature of the objects to be tracked gives more challenges for the tracker. Numerous methods have been proposed, and especially in the recent years, more tracking algorithms for relatively high-end applications are being developed. These methods can be compared based on their robustness and efficiency to tracking in different samples of videos and also on standard public datasets. In spite of all these developments, there are applications that do not need advanced and costlier systems; they rather need simple tracking systems that are lightweight, cost-effective, and rich in accuracy, speed, and real-time performance. The rest of the paper is organized as follows. In Sect. 2 we throw light precisely on some works that get along well with tracking. Section 3 presents the visual tracking system based on our proposed method. Section 4 presents the experiment results and analysis. Section 5 ends with the conclusion and future plans and scope of the research.

2 Related Work

Commonly used visual tracking methodologies have their base in the conventional mathematical concepts. However, their effective functionality and uniqueness depends on the specific environment or video scene. Frame differencing is a conventional practice of detection of object in a scene. Yongseok Yoo [4] presented an algorithm exclusively for smart cameras that is slightly better than temporal frame difference. As an advancement, Zhan Chaohui et al. [5] proposed an improved moving object detection algorithm coupling with edge detection methods. Shanshan Li et al. [6] proposed a method for moving object detection using Gaussian mixture model. Background subtraction methods [7–10] with different types of it including Gaussian, MinMax method, kernel density estimation, code books, and eigenvalues were also implemented by other authors. In the machine vision field, the mean shift was introduced by Comaniciu et al. [11] who used it for, inter alia, object tracking. Kalman filter was first described and partially presented in scientific papers by Swerling (1958) and Kalman and Bucy (1960–1961). This Kalman filtering technique was employed in visual tracking. Xin Li et al. [12] in his paper has used Kalman filter for visual tracking. A. H. Mazinan [13] proposed a new method in association with Kalman filtering approach. Particle filters are also in use for visual tracking. A. Yao et al. [14] introduced an association method of particle filtering along with kernel-based tracking. Budi Sugandi et al. [15] used a color-based particle filtering approach. Support vector machines, hidden Markov models, and neural network classifiers are used in particle filtering. Various other filters like shape filter, Bernoulli's filter, and PHD filter are also in use. Feature-based tracking approaches used color, texture, shape, etc., as necessary tracking features. Generative and discriminative methods of tracking using machine learning approaches use principal component analysis (PCA), Bayesian framework [16], and online boosting methods. Apart from all these listings, the usage of algorithms can be singular, also hybrid, i.e., fusion of different approaches to get a vastly improved result. All of these strategies depend on the type of environment, difficulty level of the scene, and the goal of the specific application.

3 The Proposed Work

3.1 Simple Algorithm for Detection and Tracking

1. Create an empty frame F_1 of an image, and then assign it to an image variable *imgTracking* which has the same dimensions of the original video.
2. Now, convert the image frame *imgTracking* to zero, i.e., "0" represents black.
3. Get the threshold image T and the binary image B from the previous frame F_1.
4. As a preprocessing operation, use Gaussian kernel G_{f1objk}, and smooth the original image from its noise.
5. Assume the processed image as S_2.
6. Do the color format conversion from RGB color model to HSV color model.
7. Find the moments Mc_{11}, Ms_{22} of the image threshold obtained to find the exact position P of the object.
8. Seamlessly track the ball or object in the specified area of the video frame.

4 Experimental Results

We implemented the algorithm for tracking using OpenCV software on Visual Studio. The program has been tested using different colored objects under required conditions. We were able to detect and track different colored objects efficiently. The experimental setup consisted of taking live video input from a sample design of conveyor and also two outdoor scenes. The simple visual tracking implementation was found successful (Fig. 1).

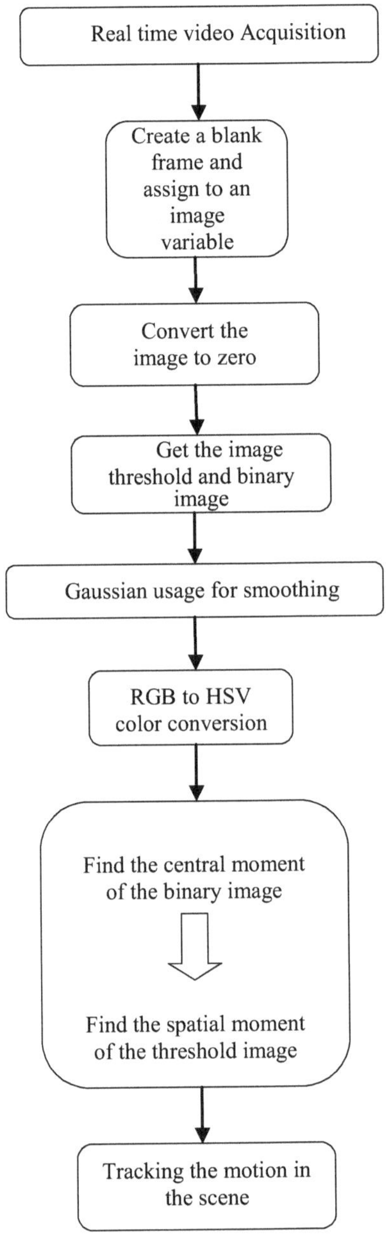

Fig. 1 Block diagram of the detection algorithm

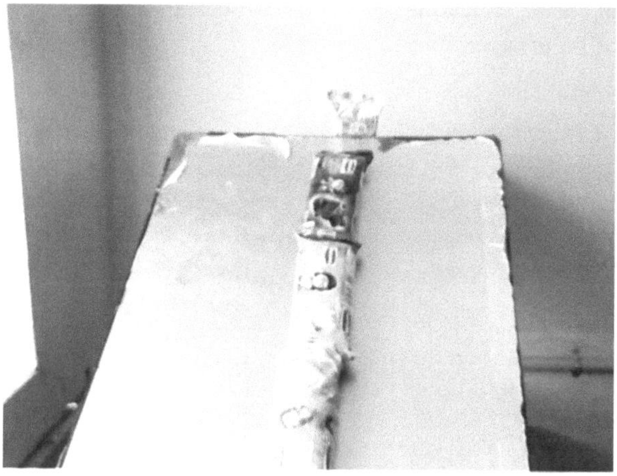

Fig. 2 The picture showing a sample of conveyor flow with objects SET 1

4.1 Result Analysis

We designed a sample setup of a conveyor system which is exclusively designed for industries. The objects are made to be released one by one, one after the other in a slope. Different objects with variety of shape, size, and color were tested under various scenarios (Figs. 2 and 3).

The process is done in such a way that whenever a specified colored object (e.g., green) passes through a slope, it gets detected and then the movement is tracked. We simply do not consider bounding box or bounding circle at the moment else a line is drawn along the motion path where the object moves, until it crosses the end of the frame. The object is passed through as much fast as possible in order to check whether the algorithm is effective, and the results proved to be the same (Figs. 4, 5, 6, and 7).

4.2 Comparison of Computational Time

Comparisons are made with mean shift tracker and the experimental results show higher detection rate. Fast-moving objects are also tracked and marked clearly that clearly indicates that this is an ideal move toward the development of tracking of fast-moving scenes in industry-oriented quality testing applications. False detections were not found even in illumination changes. However, the limitation of this method is that the count of objects to be tracked cannot be high. Hence, this is applicable for specific detection applications only (Tables 1 and 2).

Fig. 3 The picture showing SET 2 of objects

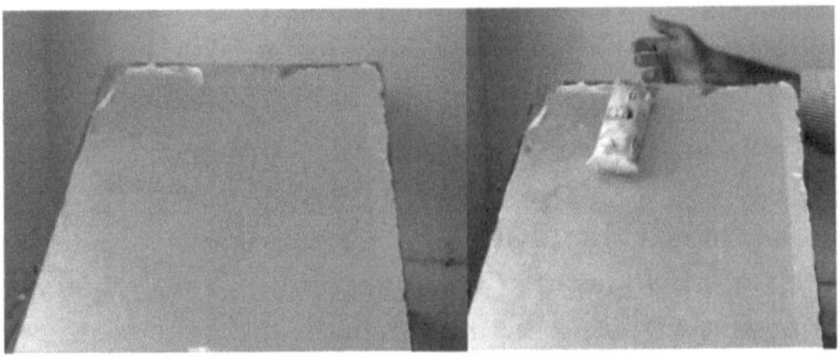

Fig. 4 The *blue marker line* denotes that a specific color packet has passed

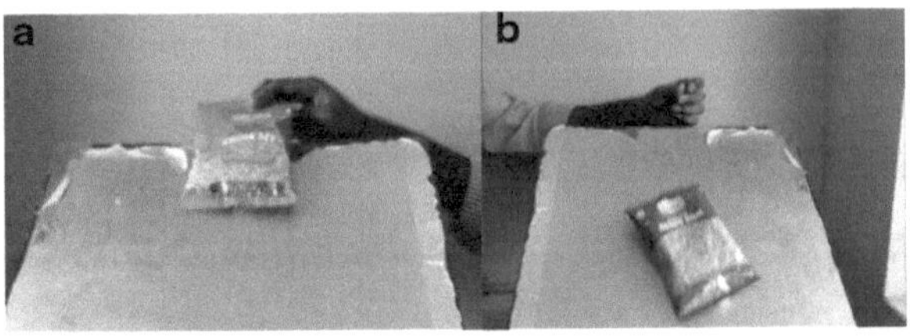

Fig. 5 Detection of a colored object (**a**) *orange* and (**b**) *red*

5 Conclusion

This paper focuses on building a simple visual tracker that can be applied in tracking unique colored objects in simple real-time scenarios only. The ultimate objective is to build a lightweight, high-performance, fast tracker that is easily affordable and more reliable. We are working toward contributing new improvements in the field of visual tracking covering a wide range of features and providing highly robust and reliable tracking for different and new industrial applications.

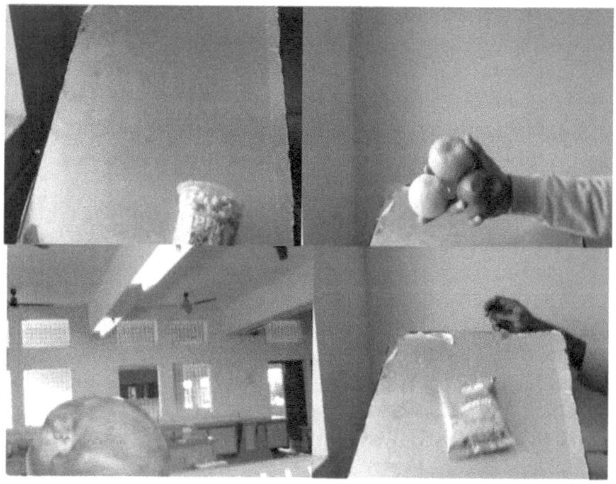

Fig. 6 Various results of the proposed tracking method

Fig. 7 Experimental results of the proposed method performing under complex backgrounds in outdoor scene

Table 1 Performance versus mean shift tracker

Time frames	Mean shift tracker (in seconds)	Proposed tracker (in seconds)
1	2.50	2.00
2	4.00	3.00
3	7.00	6.00
4	8.50	6.50

Table 2 Reduction in false alarms

Frames	False-positive rate (FP) (in percentage %)	Mismatches (in number)
1	5.75	1
2	14.5	3
3	20.5	4

References

1. Yang, H., et al.: Recent advances and trends in visual tracking. A review. Neurocomputing **74**(18), 3823–3831 (2011)
2. Benezeth, Y., Jodoin, P.M., Emile, B., Laurent, H., Rosenberger, C.: Comparative study of background subtraction algorithms. J. Electron. Imaging **19**(3), 033003 (2012)

3. Shi, J., Tomasi, C.: Good features to track. In: 1994 IEEE Computer Society Conference on Computer Vision and Pattern Recognition, 1994. Proceedings CVPR '94, pp. 593, 600, 21–23 June 1994. doi:10.1109/CVPR.1994.323794
4. Yongseok Yoo, Tae-Suh Park: A moving object detection algorithm for smart cameras. In: IEEE Computer Society Conference on Computer Vision and Pattern Recognition Workshops, 2008. CVPRW '08, pp. 1, 8, 23–28 June 2008
5. Zhan Chaohui, Duan Xiaohui, Xu Shuoyu, Song Zheng, Luo Min: An improved moving object detection algorithm based on frame difference and edge detection. In: Fourth International Conference on Image and Graphics, 2007. ICIG 2007, pp. 519, 523, 22–24 Aug 2007
6. Shanshan Li, Yuehuan Wang, Wenhui Xie, Zhiguo Cao: Moving object detection and tracking in video surveillance system. In: Proceedings of SPIE 7495, MIPPR 2009: Automatic Target Recognition and Image Analysis, 74952D, 30 Oct 2009
7. Stauffer, C., Grimson, W.E.L.: Adaptive background mixture models for real-time tracking. In: IEEE Computer Society Conference on Computer Vision and Pattern Recognition, 1999, vol. 2, pp. 252, 1999
8. Elgammal, A., Harwood, D., Davis, L.: Non-parametric model for background subtraction. In: ECCV '00 Proceedings of the 6th European Conference on Computer Vision-Part II, pp. 751–767, 2000
9. Kim, K., Chalidabhongse, T.H., Harwood, D., Davis, L.: Real-time foreground-background segmentation using codebook model. Real-Time Imaging **11**, 172–185 (2005)
10. Oliver, N.M., Rosario, B., Pentland, A.P.: A Bayesian computer vision system for modeling human interactions. Pattern Anal. Mach. Intell. **22**, 831–843 (2000)
11. Comaniciu, D., Ramesh, V., Meer, P.: Real-time tracking of non-rigid objects using mean shift. In: IEEE Conference on Computer Vision and Pattern Recognition, 2000. Proceedings, vol. 2, pp. 142, 149, 2000
12. Xin Li, Kejun Wang, Wei Wang, Yang Li: A multiple object tracking method using Kalman filter. Information and Automation (ICIA), IEEE Page(s), pp. 1862–1866. (2010)
13. Yao, A., Lin, X., Wang, G., Yu, S.: A compact association of particle filtering and kernel based object tracking. Pattern Recogn. **45**(7), 2584–2597 (2012)
14. Mazinan, A.H., Amir-Latifi, A.: A new algorithm to rigid and non-rigid object tracking in complex environments. Springer, London (2012)
15. Sugandi, B., Kim, H., Tan, J.K., Ishikawa, S.: A color-based particle filter for multiple object tracking in an outdoor environment. Artif. Life Robotics **15**(1), 41–47 (2010)
16. Serratosa, F., Alquézar, R., Amézquita, N.: A probabilistic integrated object recognition and tracking framework. Expert Syst. Appl. **39**(8), 7302–7318 (2012)

Iteration-Free Fractal Image Compression Using Pearson's Correlation Coefficient-Based Classification

M.L. Valarmathi, M. Sobia, and R. Babyla Devi

Abstract Fractal image compression (FIC) is a relatively recent technique that represents an image by a contractive transform on the space of images, in which the fixed point is close to the image. Conventional fractal image compression method requires a large amount of encoding time in range-domain mapping. In this chapter, a novel noniterative fractal compression method is proposed to meet both the quality of output image and to speed up the encoding process of FIC. The proposed method works based on the fact that the absolute value of Pearson's correlation coefficient (APCC) between two image blocks is equal to the affine similarity between them. Here, all the domain blocks are classified into three classes according to the APCC value. When matching block for a range block is searched, instead of searching the entire domain pool, only appropriate classified domain pool is used. Here, the mean image of the given image is taken as the domain pool. This proposed method can significantly speed up the encoding process as well as preserve the reconstructed image quality.

Keywords Image compression • Fractal • Domain block • Range block • Pearson's correlation coefficient • Mean image

1 Introduction

Fractal image compression and representation is based on the fact that there are tremendous similarities and redundancies in most real-world images. Fractals have detail at every scale, fractals are self-similar, fractals have dimensions, and fractals can be described by a mathematical algorithm. A fractal is a mathematical model of a self-similar nature of the real world [1]. The basic implementation of the fractal-

M.L. Valarmathi (✉) • M. Sobia • R. Babyla Devi
Department of Computer Science and Engineering, Government College of Technology, Coimbatore, TN, India
e-mail: drmlv@gct.ac.in; shobiagct@gmail.com; babyladevi@gmail.com

E.B. Rajsingh et al. (eds.), *Informatics and Communication Technologies for Societal Development*, DOI 10.1007/978-81-322-1916-3_16, © Springer India 2015

based compression process starts by taking an image and dividing it into a number of nonoverlapping blocks of the same size called range blocks. The range blocks should completely cover the entire image. Then a domain pool is constructed. Domain pool is a collection of larger overlapping image blocks from the same image. For each range block R, a domain block D that is similar to the range block is to be searched from the domain pool using contractive affine transformation [2] given by

$$\hat{R} = i\{\alpha(S \circ D) + \delta\} \tag{1}$$

where $\hat{R}$ is the matching domain block that matches with the range block after the transformation, α is the contrast scaling, S is the contraction operation that maps the domain block D to a range block, δ is the luminance shift, and i is the isometric transformation index. Then the fractal code describing the above contractive affine transformation that has minimum matching error between the original range block and the matching domain block is transmitted or stored. This fractal code consists of contrast scaling α, luminance shift δ, isometric i, and position (x, y) of the best matching domain block in the domain pool. The space required for storing the fractal code is much less than the space required for storing the image. Thus, compression is achieved. The large amount of time spent in the above fractal image compression process is because of mapping the range block by an equivalent domain block. Most pure fractal-based schemes are not competitive with the current state of the art, but hybrid schemes incorporating fractal compression and alternative techniques achieve considerably greater success [3]. Decompression is an iterative process, taking an arbitrary image as the initial image. Decoded image is repeatedly reconstructed by applying the contractive affine transformations to the iterated image.

2 Related Works

In recent years, many works had been conducted to improve the process of matching the domain block with range block in FIC [4–8]. In variance-based accelerating [9] and region-based fractal image compression schemes, only selected part of the domain blocks is taken to reduce the searching time for a range block [10]. To find the domain block that matches with the range block under consideration, exhaustive search of all the domain blocks of the image should be done. For an image of size M × M, with range block size B × B and domain block size D × D, there will be M/B × M/B range blocks. It may be noted that the next domain block is obtained by shifting the previous domain block by one pixel position. Hence, the total number of possible domain blocks in an M × M image is (M – D + 1)2. Thus, the number of domain blocks to be searched is enormous. Each of these blocks has to be tried with eight different isometric transformations plus one intensity adjustment. All this means enormous computational burden. Many fast FIC schemes are proposed to reduce the time taken for this matching process.

For example, several fast FIC schemes were developed on the basis of spatial correlation between the range and domain blocks. These fast FIC schemes speed up the matching process by searching the matching domain block from the adjacent domain block of the current range block [11] or from the adjacent domain blocks of the domain block which is the matching block for the adjacent range block of the current range block [12]. Some of the fast FIC schemes were proposed based on the classification method [1–3]. For example, Fisher's fast FIC scheme proposed in 1992 [13] is a typical classification scheme. This classification method classifies the domain blocks into some group of classes; then the matching block for a range block can be searched from the group of classified classes that closely matches with the range block. Some other schemes such as fuzzy clustering using DCT [14], Kovacs's classification [15], and sorting with some features were applied to speed up the process of encoding in fractal image compression. Although many fast schemes were proposed to speed up the process of encoding in fractal image compression, the encoding time is still too long. Hence, a fast fractal image compression method is required to speed up the process of encoding and to improve the quality of the reconstructed image.

3 Proposed Method

Conventional fractal image compression scheme requires a large amount of encoding time for performing range-domain mapping. It is decreased by classification of image parts based on Pearson's correlation coefficient [16]. The proposed method works based on the conclusion that the absolute value of Pearson's correlation coefficient (APCC) between two image blocks is equal to the affine similarity between them [17]. In the proposed method, all the domain blocks are classified into three classes according to the Pearson's correlation coefficient classification method. Moreover, the iterative decoding process is simplified by employing the iteration-free method.

3.1 *Iteration-Free Fractal Image Compression*

In conventional methods, any arbitrary image is taken as the starting image. It is iterated to get the compressed image. That is, in the decompression (to obtain the original image) process, an arbitrary initial image is partitioned into blocks. Using the fractal code for the first block, i.e., (x, y, i, α, and δ), the first block is constructed and so on up to the last block. This gives the first approximation of the required image. This approximate image is then partitioned, and the reconstruction process is carried out using this image, obtaining a still better reconstructed image. This iterative process is continued. It soon converges, yielding the original image with some small error. If a good image is taken as the reference image (instead of any arbitrary image), the blocks of the reconstructed image can be obtained at one stroke using the received fractal codes and the reference image; without any iteration Chang and Kuo [18] have adopted the mean image of the original image as the reference image transmitted along with the fractal codes to the decoder. Hence, Chang and Kuo's method is also called iteration-free method. The time taken for the iterated process is reduced by iteration-free method.

3.2 APCC-Based Fractal Image Compression

The Pearson correlation coefficient is a measure of the strength between the two linear association variables, denoted by r taking a range of values from +1 to −1.

3.3 APCC Computation

The input image of size M × M is partitioned into the nonoverlapping range blocks of size B x B, and compute the mean image which is used as the domain pool. Then APCC for all the range and domain blocks is calculated. The computational procedure for APCC is described below.

Correlation coefficient is computed by substituting the estimated value of the covariance and variances of the range and domain blocks into the formula given below. Correlation coefficient is represented by the letter "*r*." The formula for *r* is

$$r = \frac{SS_{XY}}{\sqrt{(SS_{XX})(SS_{YY})}} \tag{2}$$

where

$SS_{XX} = \sum (x_i - \overline{x})^2$ is the variance of the block *R*.

$SS_{YY} = \sum (y_i - \overline{y})^2$ is the variance of the block *D*.

$SS_{XY} = \sum (x_i - \overline{x})(y_i - \overline{y})$ is the covariance between *R* and *D*.

Based on the "*r*" values, the domain pool is grouped into the following three clusters. Let min represent the minimum of the APCC values and max represent the maximum of the APCC values.

Group the domain blocks having APCC values ranging from min to min + ¼ (max-min) as the first cluster.
Group the domain blocks having APCC values ranging from min + ¼(max-min) to min + ¾ (max-min) as the second cluster.
Group the domain blocks having APCC values ranging from min+3/4(max-min) to max as the third cluster.

During the mapping process, the APCC values of every range block are calculated. Based on this value, search is made in any one of the three domain block clusters. By this step, the total domain block number in the three domain classes is only one eighth of the domain block number in BFIC, which can greatly reduce the pair-wise comparisons between the range blocks and domain blocks. Moreover, all the domain and range blocks are classified into 3 classes to reduce the pair-wise comparisons further. More importantly, the reconstructed image quality is well preserved.

3.4 Algorithm

Steps in Encoding Process

1. Partition the input image of size M x M into the nonoverlapping range blocks of size B x B, and compute the mean image which is used as the domain pool.
2. Compute APCC for all the range and domain blocks.
3. Cluster the domain pool based on the APCC values.
4. For each range block, check if the variance of the range block is less than the threshold value (here threshold value is chosen as 5), code the range block by the mean value, and go to step 6.
5. Find the cluster to which the range block belongs to, based upon its APCC value. Then, code the range block by transformation using scaling and rotation within the respective cluster of domain blocks to find the transformation parameters that give minimum error between the transformed domain block and the range block.
6. Repeat step 4 for all range blocks. Finally, attach a header to the fractal code of each range block. It represents the coding status of the range block, i.e., whether it has been coded by mean or by transformations. Then, the mean image and the fractal codes are stored together as the representation of the compressed image.

Steps in Decoding Process

1. Retrieve the mean image from the compressed image.
2. Get the fractal code of the first range block. Use the header information to code the corresponding range using the mean or by transformations.

    2.1 If the range block is coded by mean, then fill corresponding block with the mean value of the range block obtained from the mean image.

2.2 If the block is to be coded by transformations, apply scaling and rotations to the position of the mean image as in the fractal code. Fill the range block in the blank image with the transformed block.

4 Implementation and Experimental Results

The proposed method is tested using 256 × 256 grayscale images. The size of range block is set to be 4 × 4, and the search step size $\delta h = 4$. To compare the performance, the following parameters are taken into account.

(i) Compression time
(ii) Decompression time
(iii) Compression ratio
(iv) PSNR (peak signal-to-noise ratio)

The performance of the decoded image is evaluated by the PSNR. The distortion between the original and coded range blocks is represented by the mean square error (MSE) measurement. The amount of compression is calculated by compression ratio. The timing complexity is measured by the amount of time required to encode and decode the image. It is measured in seconds(s).

5 Results and Discussion

The simulation is done in Intel core i7 CPU with 3.40 GHz processing speed and with 2.00 GB RAM. Standard image like Mandrill, Dune, Lena, Parrots, and Pepper are tested. The comparison of compression time for proposed and iteration-free method is shown in Fig. 1. The comparison of PSNR value for proposed and iteration-free method is shown in Fig. 2. Mean image and classified blocks for the test image using Pearson's correlation coefficient-based fractal image compression are shown in Fig. 3b, c. The compression and decompression are done for various test images using APCC-based fractal image compression with 4 × 4 block size, and the resulting images are shown in Fig. 4.

Table 1 shows the comparison between the compression by iteration-free method and APCC method for grayscale images. Compression time for the images in iteration-free method and APCC is given. The comparative reduction of compression time is calculated. The reduction in compression time is given as compression time reduction factor. There is a huge reduction in the compression time in APCC method than iteration-free method. The compression time reduction factor varies with the test images. From the table, it shows the results of iteration-free method with step size 1 and APCC method with step size 4. The maximum compression time reduction factor is 43.35; it occurs for the Lena image. The minimum compression time reduction factor is for the Dune image which is

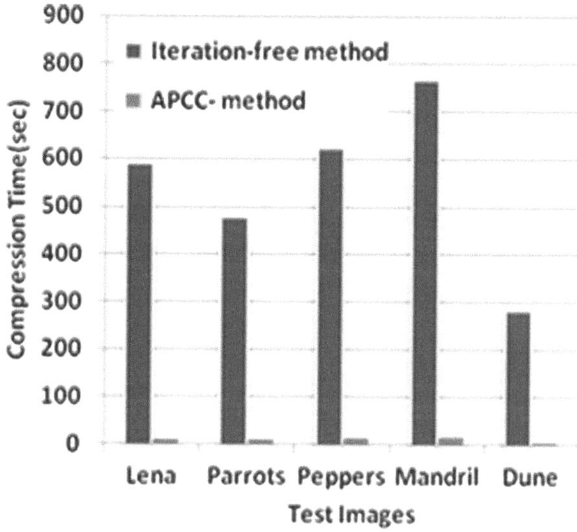

Fig. 1 Compression time for proposed and iteration-free method

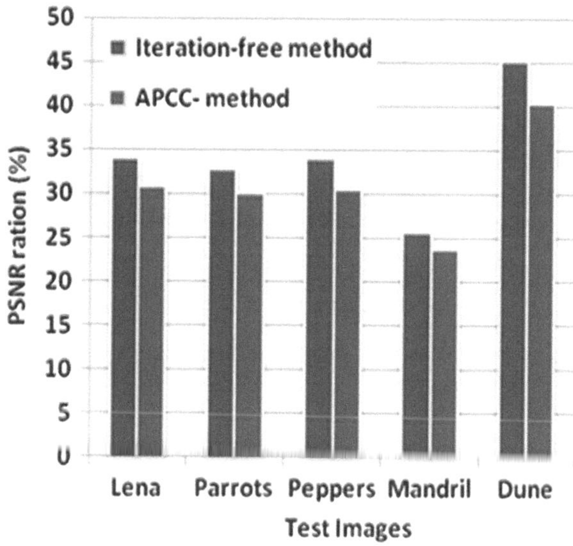

Fig. 2 PSNR value for proposed and iteration-free method

about 43.10. The average compression reduction factor is about 43.26; this means the compression time in APCC method is extremely low when compared to that of base iteration-free method. However, by using APCC there is a slight reduction in the quality of the image.

The amount of reduction in quality is given by PSNR reduction. This PSNR reduction is variable with the test images. The highest value of reduction in quality of reconstructed image by APCC method is for the Dune image. It has the PSNR reduction of 10.57 %. The minimum reduction in quality is for the Mandrill image.

Fig. 3 Outputs for the mean image and classified blocks

Fig. 4 Comparison of original and reconstructed images

Table 1 Test results for APCC-based method and iteration-free method

	Compression ratio (%)	Compression time (sec)		Compression time reduction factor	PSNR		PSNR reduction in (%)
Name of the image	Iteration-free and APCC methods	Iteration-free method	APCC method		Iteration-free method	APCC method	
Lena	82.64	588.26	13.57	43.35	33.98	30.81	9.33
Parrots	84.59	478.27	11.05	43.28	32.81	30.00	8.56
Peppers	82.05	621.66	14.37	43.26	33.98	30.42	10.48
Mandrill	79.46	767.40	17.72	43.31	25.65	23.61	7.95
Dune	88.07	282.31	6.55	43.10	45.12	40.35	10.57

It has PSNR reduction of 7.95 %. The overall reduction in the quality is very low. The average PSNR reduction is about 9.38 %. Hence, the APCC method with step size four tremendously reduces the compression time when compared to the base iteration-free method with step size one, with an acceptable amount of quality.

6 Conclusion and Further Extension

The long encoding time in fractal image compression is one of the major difficulties for its application. Although several researches were proposed to improve the process of encoding in FIC, still the time taken for encoding is too long, and in many proposed schemes, the quality of the reconstructed image is not in the acceptable format. Based on the fact that the affine similarity between two image blocks in FIC is equivalent to the absolute value of Pearson's correlation coefficient between them, a new fast FIC scheme is proposed. In this proposed system, all the domain blocks are classified into three classes using Pearson's correlation coefficient method. Moreover, the iterative decoding process is also simplified by employing the iteration-free method. In this method there is no iteration since the same mean image of the original image is used as domain blocks in encoding and decoding. The experiment results show that the proposed scheme can greatly reduce the encoding time with preserving the reconstructed image quality well.

As a future enhancement to this chapter, classification schemes can be significantly enhanced using several optimization algorithms to achieve the better image quality and faster coding time. This method of image compression can be extended to color images in the future.

References

1. Fisher, Y.: Fractal Image Compression: Theory and Application. Springer-Verlag, New York (1995)
2. Jacquin, A.E.: Fractal image coding: a review. Proc. IEEE **81**, 1451–1465 (1993)
3. Wohlberg, B., de Jager, G.: A review of the fractal image coding literature. IEEE Trans. Image Process. **8**(12), 1716–1729 (1999)
4. Schwartz, W.R., Pedrini, H.: Improved fractal image compression based on robust feature descriptors. Int. J. Image Graph. **11**(4), 571–587 (2011)
5. Chen, H.N., Chung, K.L., Hung, J.E.: Novel fractal image encoding algorithm using normalized one-norm and kick-out condition. Image Vis. Comput. **28**(3), 518–525 (2010)
6. Zhou, Y., Zhang, C., Zhang, Z.: An efficient fractal image coding algorithm using unified feature and DCT. Chaos Soliton. Fract. **39**(4), 1823–1830 (2009)
7. Distasi, R., Nappi, M., Riccio, D.: A range/domain approximation error-based approach for fractal image compression. IEEE Trans. Image Process. **15**(1), 89–97 (2006)
8. He, C., Xu, X., Li, G.: Improvement of fast algorithm based on correlation coefficients for fractal image encoding. Comput. Simul. **12**(4), 60–63 (2005)

9. He, C., Yang, S., Huang, X.: Variance-based accelerating scheme for fractal image encoding. Electron. Lett. **40**(2), 1052–1053 (2004)
10. Hartenstein, H., Ruhl, M., Saupe, D.: Region-based fractal image compression. IEEE Trans. Image Process. **9**(7), 1171–1184 (2000)
11. Truong, T.K., Kung, C.M., Jeng, J.H., Hsieh, M.L.: Fast fractal image compression using spatial correlation. Chaos Soliton. Fract. **22**(5), 1071–1076 (2004)
12. Wang, J., Liu, Y., Wei, P., Tian, Z., Li, Y., Zheng, N.: Fractal Image Coding Using SSIM: In Proceedings of 18th ICIP, pp. 245–248 (2011)
13. Fisher, Y., Jacobs, E.W., Boss, R.D.: Fractal image compression using iterated transforms. Image Text Compress. **176**, 35–61 (1992)
14. Jaferzadeh, K., Kiani, K., Mozaffari, S.: Acceleration of fractal image compression using fuzzy clustering and discrete-cosine-transform-based metric. IET Image Process. **6**(7), 1024–1030 (2010)
15. Kovacs, T.: A fast classification based method for fractal image encoding. Image Vis. Comput. **26**(8), 1129–1136 (2008)
16. Wang, J., Zheng, N.: A novel fractal image compression scheme with block classification and sorting based on Pearson's correlation coefficient. IEEE Trans. Image Process. **22**(9), 3690–3702 (2013)
17. Wang, X., Wang, Y., Yun, J.: An improved fast fractal image compression using spatial texture correlation. Chin. Phys. B **20**(10), 104202-1–104202-11 (2011)
18. Chang, H.T., Kuo, C.J.: Iteration – free fractal image coding based on efficient domain pool design. IEEE Trans. Image Process. **9**(3), 329–339 (2000)

A Resilient Digital Watermarking Exploiting Spatial and Frequency Domain Against Image Hackers

R. Kalpanasonika and S. Akila Agnes

Abstract The process of image hacking has been amplified these days due to the advent of the Internet and computer technology in this era. A fortunate thing in digital media is digital watermarking which endows various benefits such as copy control, authentication and rights management of digital media. The aim of this work is to improve the secure transmission and proof of ownership which are the two significant applications of the digital image watermarking, and it is achieved by combining the frequency domain (three-level discrete wavelet transforms and scrambling process) along with the spatial domain technique (LSB substitution technique in CbCr colour space and additive pixel technique) in order to improve the visual quality of a watermarked image.

Keywords Digital watermarking • LSB substitution • Frequency domain • Chaotic map • Discrete wavelet transforms • Colour spaces • Spatial domain

1 Introduction

Digital media techniques are the foremost driving force behind the emergence of the Internet. To protect images from a malicious user and to provide evidence for ownership, there are several methods, but the best approach for digital image protection is the watermarking concept. Digital watermarking is the process of embedding any multimedia content into a cover source which indicates the provision of ownership to the authorized person. This digital watermarking process can be done either by exploiting the spatial domain or the transform domain. But in this proposed concept, the digital image is watermarked by combining both techniques.

R. Kalpanasonika (✉) • S. Akila Agnes
Department of Computer Science and Engineering, Karunya University, Coimbatore, TN, India
e-mail: kalpanasonika@gmail.com; akilaagnes@karunya.edu

E.B. Rajsingh et al. (eds.), *Informatics and Communication Technologies for Societal Development*, DOI 10.1007/978-81-322-1916-3_17, © Springer India 2015

Though the spatial domain technique is easy to implement with low complexity, it is fragile to geometric attacks in image processing, so specifically the spatial domain uses LSB substitution in the chroma component of the image to be highly secured for the embedding process. In the frequency domain, it uses two transform models, the three-level discrete wavelet transform and Arnold's cat map transform (ACM) [1–5], which have good embedding capability and robustness when compared to the spatial domain techniques. Discrete wavelet transform (DWT) [4, 6, 7] is one of the prevailing wavelet techniques in the frequency domain with minimal computation complexity rate. The DWT decomposes an image into various sub-band levels which include four different sub-band levels low to low (LL), low to high (LH), high to low (HL) and high to high (HH), correspondingly. The low to low sub-band of the first level gets decomposed to get the second decomposition level. The low to low sub-band of the second level is further decomposed to the third level of the discrete wavelet transform.

2 Proposed Method

The proposed concept securely embeds the watermark logo and watermark text in the cover image with the usage of the spatial and transform techniques. This comprises two major phases: the embedding and extraction phases.

2.1 Embedding Course of Action

- Any colour image of size 512 × 512 pixels is used as cover image.
- The three-level decomposition of DWT is done on the cover image in separate channels.
- The watermark logo of 50 × 50 pixels gets resized to 512 × 512 pixels of the three-level decomposed cover image.
- The resized logo gets scrambled [1–5] with the help of Arnold's cat map transform model.
- A spatial domain technique named additive pixel is performed by combining the scrambled logo and the three-level decomposed cover image of 512 × 512 pixels.
- The resultant image performs inverse 3DWT [4, 6, 7] to regain back its 512 × 512 pixel size of the logo-embedded image (Fig. 2).
- The resized resultant image gets converted into YCbCr colour image, and then the watermark text is embedded by converting the watermark text into binary form.
- In the YCbCr colour image, the Y component is not to be considered for embedding because the human eye is more receptive to brightness and less to the colour difference of Cb and Cr chroma components.

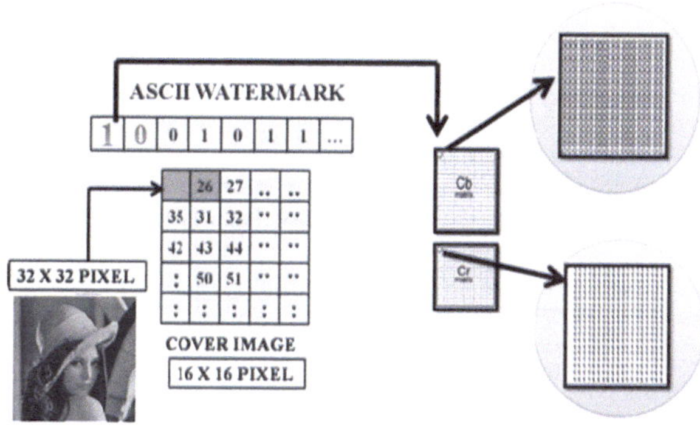

Fig. 1 Block splitting process

- The watermarked image 1 is converted into a 2D matrix, which gets segmented into blocks and for embedding text in Cb Cr channel it is done with some conditions like for every even block position of the watermarked Image 1, if the watermark text bit value is 1 then the LSB [8] of Cb is set to 0 and Cr as 1 for the entire block. If the watermark text bit value is 0 then the LSB of Cb is set to 1 and Cr as 0 for the entire block (Fig. 1).
- For every odd block position of the watermarked Image 1, if the watermark text bit value is 1 then the LSB of Cb is set to 1 and Cr as 0 for the entire block. if the watermark text bit value is 0 then the LSB of Cb is set to 0 and Cr as 1 for the entire block. Finally, the watermarked text and logo-embedded image gets converted from YCbCr to an RGB-coloured image (Fig. 2).

2.2 Extraction Course of Action

- To extract the watermark text [8] from the watermarked image, the watermarked image should be converted to YCbCr colour space image, and in Cb block, if the occurrences of 0 are more when compared to that of 1, and in Cr if the occurrences of 1 are more than 0, then the watermark binary value is set to be 1. If this condition is not satisfied, then the watermark binary value is set to 0 for even pixel.
- In the case of odd pixel position, in Cb if the occurrences of 1 are more when compared to 0 and in Cr if the occurrences of 0 are more when compared to 1, then the watermark binary value is set to 1. If this condition is not satisfied, then the watermark binary value is set to 0.
- After the extraction of watermark text, the resultant image which contains the embedded scrambled logo gets separated by decomposing it to 3DWT. The logo-embedded 3DWT of the cover image is subtracted with the 3DWT of the cover image to produce the scrambled logo, and finally, inverse scrambling is

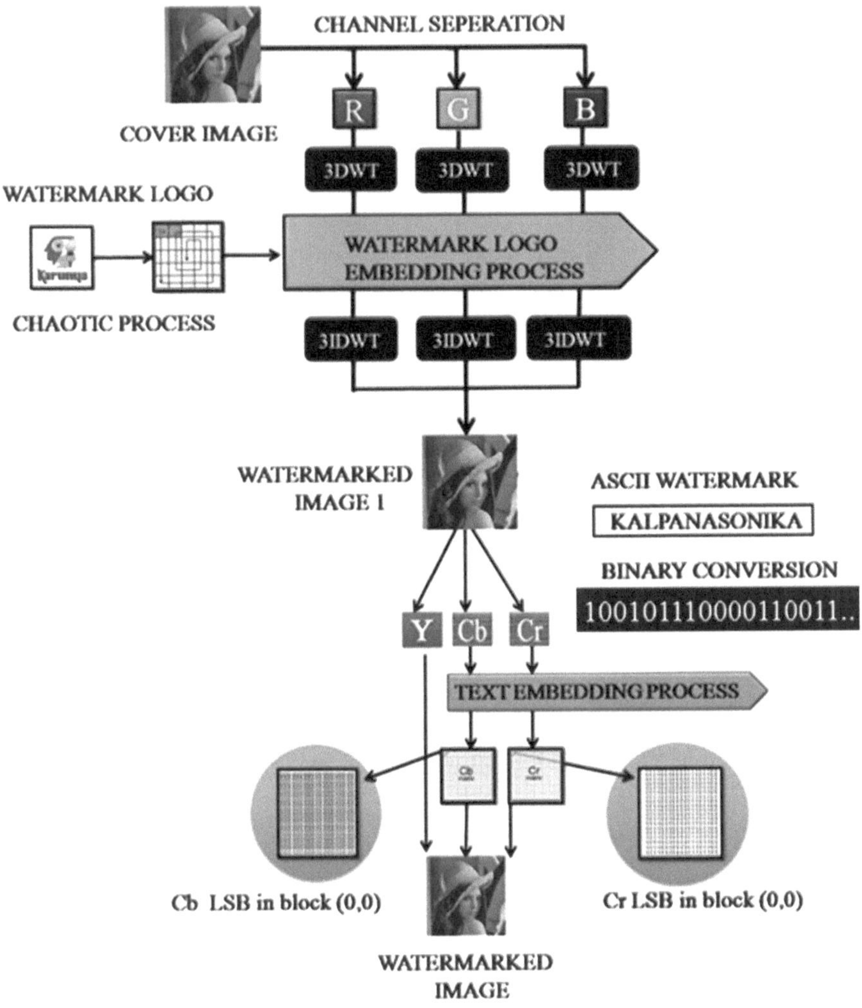

Fig. 2 Embedding system architecture

performed and it is resized to 50 × 50 pixels to regain the original watermark logo (Fig. 3).

3 Experimental Result and Discussion

This segment will illustrate the performance of the proposed concept with experimental results. The test images “Lena”, “Pepper”, “Crown”, “Mandrill”, “Plane” and “Boat” are used as cover images which are of 512 × 512 pixels. Here the

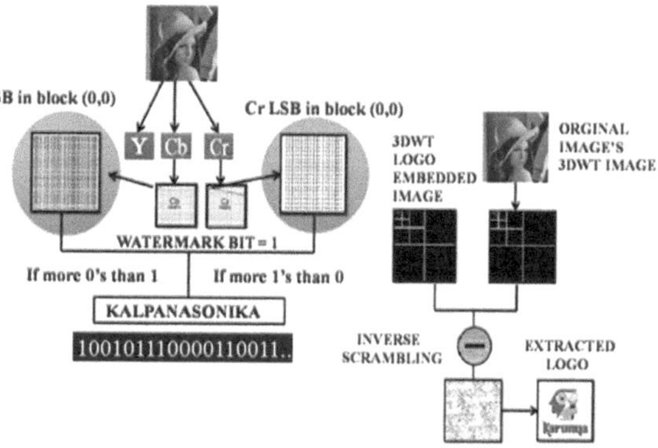

Fig. 3 Watermark extraction process

performance gets calculated on quality and visibility basis. On visibility basis, PSNR (peak signal-to-noise ratio), MSE (mean square error) and SSIM (structural similarity index method) are calculated. SSIM metric is used to improve the traditional methods like PSNR and MSE quality according to human eye perception.

If the SSIM value is 1, then the two images compared are indistinguishable, and if the PSNR value is greater than 35 db, it means that the two images which are evaluated are not perceptibly different in human vision (Table 1).

Here the embedded image in RGB colour channel, YCbCr colour channel and hybrid method is compared with the proposed method. Even though the **PSNR** value is above 35 db for the above-mentioned methods, the proposed method has **42.6904 db** as higher value for the test image (Boat). From this, we can infer that the embedded image is not visibly different from the original image.

4 Conclusion and Future Work

A systematic digital watermarking concept should satisfy security, robustness and imperceptibility for an image to be steadily watermarked; hence, no hackers can sense or confiscate watermark from the watermarked image until the commercial value of the image gets degraded. As per the experimental results, the visibility measure of the watermarked image and cover image has the maximum **PSNR** value of **42.6904** db and the **SSIM** value of **0.999761** as its utmost value. Even though we could retrieve the watermark content from the image, the original logo and the logo after retrieval from the watermarked image will slightly differ because of chaotic processing and change in the LSB value in Cb Cr channels. So this work can be extended to produce a visibly similar logo while extraction.

Table 1 **PSNR, MSE** and **SSIM** values of the test images for various methods

Image	Methods	PSNR	MSE	SSIM
	RGB	40.2364	6.15799	0.999761
	YCbCr	45.2298	1.71211	0.998822
	Hybrid	39.1795	7.85466	0.999721
Lena	Proposed	**40.023**	6.46819	**0.998691**
	RGB	40.2353	6.15962	0.99984
	YCbCr	45.2298	1.71211	0.998909
	Hybrid	39.2083	7.80271	0.999806
Pepper	Proposed	**40.0334**	6.45267	**0.999113**
	RGB	40.2845	6.09024	0.999884
	YCbCr	45.2298	1.71211	0.998878
	Hybrid	39.0575	8.07855	0.999866
Crown	Proposed	**40.0751**	6.39098	**0.999366**
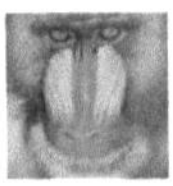	RGB	40.2374	6.15664	0.999719
	YCbCr	45.2298	1.71211	0.999517
	Hybrid	39.1515	7.90543	0.999657
Mandrill	Proposed	**40.0172**	6.47675	**0.998362**
	RGB	40.6907	5.54645	0.9995
	YCbCr	40.2348	6.12231	0.981171
	Hybrid	39.207	7.80505	0.999313
Plane	Proposed	**40.1171**	6.32951	**0.997742**
	RGB	45.3216	1.90949	0.999931
	YCbCr	40.2348	6.12231	0.998067
	Hybrid	39.1515	7.90543	0.999896
Boat	Proposed	**42.6904**	3.49977	**0.999761**

References

1. Gunjal, B.L., Mali, S.N.: Strongly robust and highly secured DWT-SVD based color image watermarking: embedding data in all Y, U, V color spaces. I.J. Inf. Technol. Comput. Sci. **4**(3), 1–7 (2012)
2. Min-Jen Tsai: Wavelet tree based digital image watermarking *by* adopting the chaotic system for security enhancement. Multimedia Tools Appl. **52**(2–3), 347–367 (2011)
3. Peng Zhua, Fei Jiab, Junliang Zhang: A copyright protection watermarking algorithm for remote sensing image based on binary image watermark. Int. J. Light Electron Opt. **124**(20), 4177–4181 (2013)
4. Shang-Lin Hsieh: An image authentication scheme based on digital watermarking and image secret sharing. Multimedia Tools Appl. **52**, 597–619 (2011)
5. Zhengjun Liu, Yu Zhang, Wei Liu, Fanyi Meng, Qun Wu, Shutian Liu: A mixed scrambling operation for hiding image. Int. J. Light Electron Opt. **124**(22), 5391–5396 (2013)
6. Dayalin Leena, G., Selva Dhayanithy, S.: Robust image watermarking in frequency domain. Int. J. Innov. Appl. Stud. **2**(4), 582–587 (2013). ISSN 2028–9324
7. Mingli Zhang, Qiang Zhang, Changjun Zhou: Robust digital image watermarking in DWT-SVD domain. In: Artificial Intelligence and Computational Intelligence. Lecture Notes in Computer Science, vol. 7003, pp. 75–84. Springer, Berlin/Heidelberg (2011). ISSN 978-3-642-23887-1
8. Lussona, F., Bailey, K., Leeney, M., Curran, K.: A novel approach to digital watermarking, exploiting colour spaces. Signal Process. **93**(5), 1268–1294 (2013)

Index

E.B. Rajsingh et al. (eds.), *Informatics and Communication Technologies for Societal Development*, DOI 10.1007/978-81-322-1916-3,